IUMRS-ICA 2008 Symposium, Sessions X and Y

Journal of Physics: Conference Series Volume 184

Nagoya, Japan
9-13 December 2008

ISBN: 978-1-61738-300-7

Printed from e-media with permission by:

Curran Associates, Inc.
57 Morehouse Lane
Red Hook, NY 12571
www.proceedings.com

Some format issues inherent in the e-media version may also appear in this print version.

Printed by Curran Associates, Inc. (2010)

Institute of Physics
Dirac House, Temple Back
Bristol BS1 6BE UK

Phone: 44 1 17 929 7481
Fax: 44 1 17 920 0979

techtracking@iop.org

TABLE OF CONTENTS

SESSION X. APPLICATIONS OF SYNCHROTRON RADIATION AND NEUTRON BEAM TO SOFT MATTER SCIENCE

SESSION Y. FRONTIER OF POLYMERIC NANO-SOFT-MATERIALS –PRECISION POLYMER SYNTHESIS, SELF-ASSEMBLING AND THEIR FUNCTIONALIZATION

PREFACE

Applications of Synchrotron Radiation and Neutron Beam to Soft Matter Science (Symposium X of IUMRS-ICA2008)

Toshiji Kanaya, Kohji Tashiro, Kazuo Sakura Keiji Tanaka, Sono Sasaki, Naoya Torikai, Moonhor Ree, Kookheon Char, Charles C Han, Atsushi Takahara

This volume contains peer-reviewed invited and contributed papers that were presented in Symposium X 'Applications of Synchrotron Radiation and Neutron Beam to Soft Matter Science' at the IUMRS International Conference in Asia 2008 (IUMRS-ICA 2008), which was held on 9–13 December 2008, at Nagoya Congress Center, Nagoya, Japan.

Structure analyses of soft materials based on synchrotron radiation (SR) and neutron beam have been developed steadily. Small-angle scattering and wide-angle diffraction techniques clarified the higher-order structure as well as time dependence of structure development such as crystallization and microphase-separation. On the other hand, reflectivity, grazing-incidence scattering and diffraction techniques revealed the surface and interface structural features of soft materials. From the viewpoint of strong interests on the development of SR and neutron beam techniques for soft materials, the objective of this symposium is to provide an interdisciplinary forum for the discussion of recent advances in research, development, and applications of SR and neutron beams to soft matter science. In this symposium, 21 oral papers containing 16 invited papers and 14 poster papers from China, India, Korea, Taiwan, and Japan were presented during the three-day symposium. As a result of the review of poster and oral presentations of young scientists by symposium chairs, Dr Kummetha Raghunatha Reddy (Toyota Technological Institute) received the IUMRS-ICA 2008 Young Researcher Award.

We are grateful to all invited speakers and many participants for valuable contributions and active discussions.

Organizing committee of Symposium (IUMRS-ICA 2008)

Professor Toshiji Kanaya (Kyoto University)
Professor Kohji Tashiro (Toyota Technological Institute)
Professor Kazuo Sakurai(Kitakyushu University)
Professor Keiji Tanaka (Kyushu University)
Dr Sono Sasaki (JASRI/Spring-8)
Professor Naoya Torikai (KENS)

Professor Moonhor Ree (POSTECH)
Professor Kookheon Char (Seoul National University)
Professor Charles C Han (CAS)
Professor Atsushi Takahara(Kyushu University)

Frontier of Polymeric Nano-Soft-Materials, Precision Polymer Synthesis, Self-assembling and Their Functionalization (Symposium Y of IUMRS-ICA2008)

Seiichi Kawahara, Rong-Ming Ho, Hiroshi Jinnai, Masami Kamigaito, Takashi Miyata, Hiroshi Morita, Hideyuki Otsuka, Daewon Sohn, Keiji Tanaka

It is our great pleasure and honor to publish peer-reviewed papers, presented in Symposium Y 'Frontier of Polymeric Nano-Soft-Materials Precision Polymer Synthesis, Self-assembling and Their Functionalization' at the International Union of Materials Research Societies International Conference in Asia 2008 (IUMRS-ICA2008), which was held on 9–13 December 2008, at Nagoya Congress Center, Nagoya, Japan.

'Polymeric nano-soft-materials' are novel outcomes based on a recent innovative evolution in polymer science, i.e. precision polymer synthesis, self-assembling and functionalization of multi-component systems. The materials are expected to exhibit specific functions and unique properties due to their hierarchic morphologies brought either by naturally-generated ordering or by artificial manipulation of the systems, e.g., crystallization and phase-separation. The emerging precision synthesis has brought out new types of polymers with well-controlled primary structures. Furthermore, the surface and interface of the material are recognized to play an important role in the outstanding mechanical, electrical and optical properties, which are required for medical and engineering applications. In order to understand structure-property relationships in the nano-soft-materials, it is indispensable to develop novel characterization techniques.

Symposium Y aimed to provide recent advances in polymer synthesis, self-assembling processes and morphologies, and functionalization of nano-soft-materials in order to initiate mutual and collaborative research interest that is essential to develop revolutionarily new nano-soft-materials in the decades ahead.

Four Keynote lectures, 15 invited talks and 30 posters presented important new discoveries in polymeric nano-soft-materials, precision polymer synthesis, self-assembling and their functionalization. As for the precision polymer synthesis, the latest results were provided for studies on synthesis of polyrotaxane with movable graft chains, organic-inorganic hybridization of polymers, supra-molecular coordination assembly of conjugated polymers, precision polymerization of adamantane-containing monomers, production of high density polymer brush and synthesis of rod coil type polymer. The state-of-the-art results were introduced for the formation of nano-helical-structure of block copolymer containing asymmetric carbon atoms, self-assembling of block copolymers under the electric field, self-assembling of liquid crystalline elastomers, preparation of nano cylinder template films and mesoscopic simulation of phase transition of polymers and so forth. Moreover, recent advantages of three-dimensional electron microtomography and scanning force microscopy were proposed for analyses of nano-structures and properties of polymeric multi-component systems. Syntheses, properties and functions of slide-ring-gel, organic-inorganic hybrid hydrogels, hydrogel nano-particles, liquid-crystalline gels, the self-oscillating gels, and double network gels attracted participants' attention. Modifications of naturally occurring polymeric materials with supercritical carbon dioxide were introduced as a novel technology.

Some of the attractive topics are presented in this issue.

We are grateful to all the speakers and participants for valuable contributions and active discussions.

Organizing committee of Symposium Y (IUMRS-ICA 2008)

Chair
Seiichi Kawahara (Nagaoka University of Technology, Japan)

Vice Chairs
Rong-Ming Ho (National Tsing Hua University, Taiwan)
Hiroshi Jinnai (Kyoto Institute of Technology, Japan)
Masami Kamigaito (Nagoya University, Japan)
Takashi Miyata (Kansai University, Japan)
Hiroshi Morita (National Institute of Advanced Industrial Science and Technology, Japan)
Hideyuki Otsuka (Kyushu University, Japan)
Daewon Sohn (Hanyang University, Korea)
Keiji Tanaka (Kyushu University, Japan)

All papers published in this volume of *Journal of Physics: Conference Series* have been peer reviewed through processes administered by the proceedings Editors. Reviews were conducted by expert referees to the professional and scientific standards expected of a proceedings journal published by IOP Publishing.

Structural Evolution Process of *isotactic* Polypropylene in the Isothermal Crystallization from the Melt

Kummetha Raghunatha Reddy[1], Kohji Tashiro[1,4], Takashi Sakurai[2], Noburu Yamaguchi[2], Makoto Hanesaka[1], Sono Sasaki[3], Hiroyasu Masunaga[3], and Masaki Takata[3]

[1] Department of Future Industry-Oriented Basic Science and Materials, Graduate School of Engineering, Toyota Technological Institute, Tempaku, Nagoya 468-8511, Japan

[2] Petrochemical Research Laboratory, Sumitomo Chemical Co. Ltd., Chiba 299-0295, Japan

[3] Japan Synchrotron Radiation Research Institute (JASRI), Kouto, Sayo-gun, Hyogo, 679-5198, Japan

E mail: ktashiro@toyota-ti.ac.jp

Abstract. The melt-isothermal crystallization behaviour has been investigated for *isotactic* polypropylene (iPP) by means of time-resolved simultaneous measurements of small-angle X-ray scattering (SAXS) and wide-angle X-ray diffraction (WAXD) using synchrotron radiation source in SPring-8. The time-resolved Fourier-transform infrared spectral (FTIR) measurements were also performed to clarify the formation process of helices. The SAXS data were analyzed on the basis of Guinier (isolated domains) and Debye-Bueche (correlated domains) theories and correlation function (stacked lamellae). The results were combined with the FTIR data, allowing us to draw the concrete structural evolution process including the formation of regular helices and their growth, the correlation of higher density domains consisting of these helices, and the increment of correlation between the neighbouring lamellae to build up the stacked lamellar structure of higher crystallinity.

1. Introduction

When a semicrystalline polymer is cooled from the melt it exhibits higher-order structure, which is built up with a complex combination of crystalline, amorphous and intermediate regions. The mechanical properties sensitively depend on this higher-order structure [1]. We need to understand the formation of higher-order structure at molecular level in the crystallization process to develop polymer materials with excellent properties. *isotactic* Polypropylene (iPP) is one of the industrially-important polymers. The crystallization behavior of iPP is quite complex. Although there have been published many reports concerning the structural evolution process of iPP, the details have not yet been clarified enough well [2-4]. In order to establish the structure-property relationship of iPP we need to understand the crystallization phenomenon from the formation of higher order structure to the regular helical chain. The combination of SAXS and WAXD techniques with the vibrational spectroscopy is one of the most useful methods for this purpose. The analysis of infrared bands with different critical sequence lengths or different number of monomeric units (m) included in the regular helix is useful to know the growth of regular helical chain [5-7]. The infrared band can be detected for the first time when the regular helical chain length exceeds the m value. The time-resolved FTIR measurements may tell us the behavior of the infrared bands with the different m values, from which the regularization process of random coils in the melt can be traced concretely [6, 7]. On the other hand, the analysis of the time-resolved SAXS and WAXD data may give us the information about the formation of stacked lamellar structure and crystal lattice, respectively. The combination of these different information obtained by IR and X-ray measurements should be quite useful to propose the structural evolution process of iPP in the isothermal crystallization phenomenon from the melt, as will be described in the present paper.

2. Experimental

The pure deuterated and hydrogenated iPP sample had the molecular weight $M_w = 1.6 \times 10^5$ g mol^{-1} and $M_w/M_n = 3.0$. In the isothermal crystallization experiment, after keeping the sample in the melt for a while to erase all the thermal histories, the sample was rapidly transferred to T_c (127 °C) by using a temperature jump cell. The simultaneous SAXS and WAXD measurements were performed in the BL40B2 beam-line at SPring-8, Japan. The time-resolved FTIR measurements were performed at time interval 3 sec using a Varian FTS7000 spectrometer.

3. Results and Discussion

Figure 1 (a) shows the SAXS profiles detected in the isothermal crystallization process after the temperature jump. Before the appearance of the so-called long period peak, the SAXS profiles change remarkably in the low q range, where q is a scattering vector. The following process might be

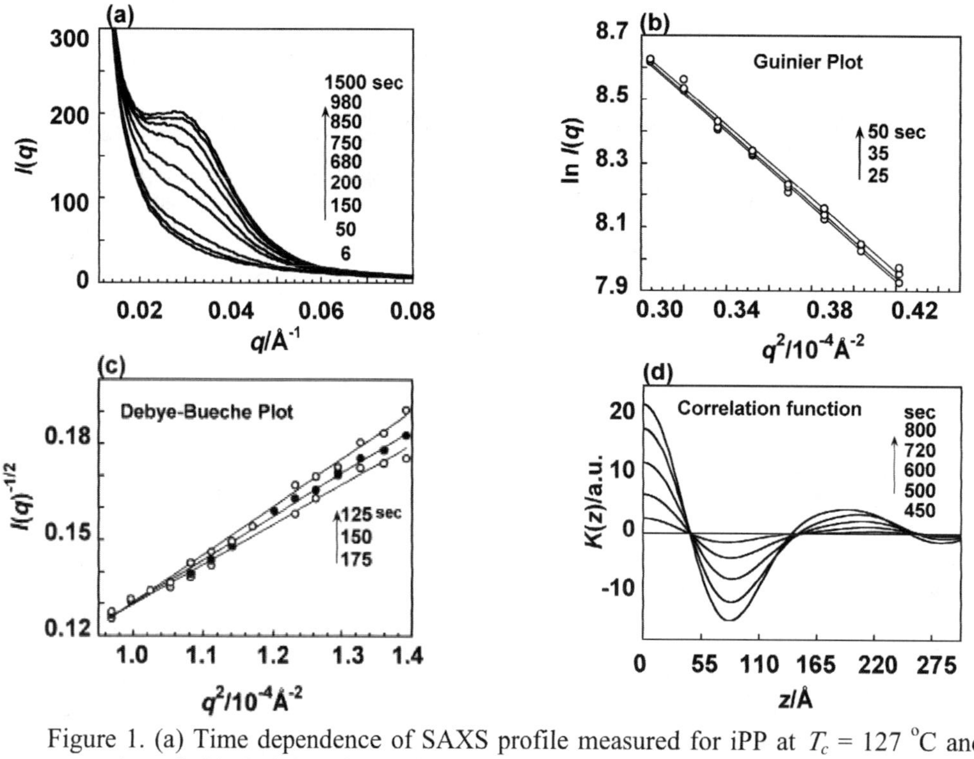

Figure 1. (a) Time dependence of SAXS profile measured for iPP at T_c = 127 °C and examples of (b) Guinier plot and (c) Debye-Bueche plots and (d) one-dimensional electron density correlation function $K(z)$ of iPP sample.

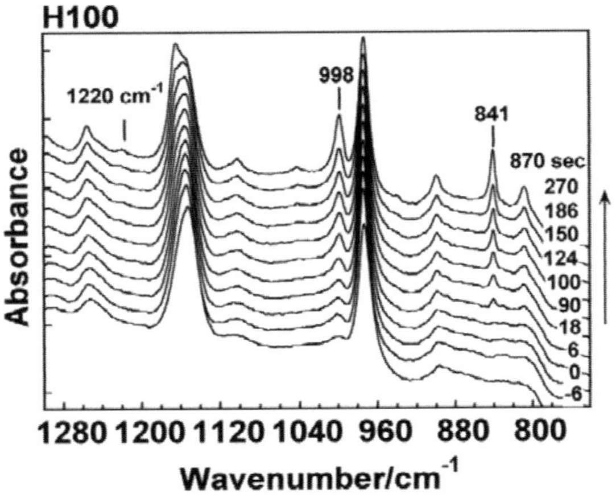

Figure 2. Time dependence of infrared spectra measured for iPP sample at T_c= 127 °C in the isothermal crystallization process from the melt.

speculated as a possibility to interpret these SAXS data. The density fluctuation may occur in the melt and some denser regions start to grow to some domains. But these domains are isolated each other. With the passage of time the domains become larger and go to gathering together, and finally a stacked lamellar structure may be formed from these highly-correlated domains. On the basis of these concepts, the SAXS data were analyzed using (i) Guinier plot for an isolated domain [$\ln[I(q)] = \ln[I(0)] - q^2R_g^2/3$, R_g = the radius of gyration of an isolated domain], (ii) Debye-Bueche plots for the two-phase structure $I(q) = A/(1 + \xi^2q^2)^2$, ξ = correlation length of neighbouring domains] and (iii) the correlation function for a stacked lamellar structure

$$[K(z) = \langle[\eta(z') - \langle\eta\rangle][\eta(z+z') - \langle\eta\rangle]\rangle = 2\int_0^{\infty} (\pi)^{-1}q^2I(q)\cos(qz)\,dq].$$

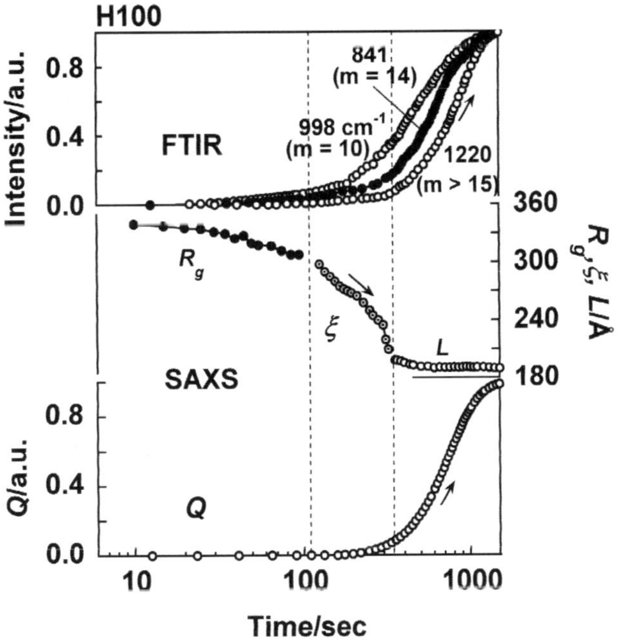

Figure 3. Comparison of infrared band intensity changes with the SAXS (invariant Q), the domain size (R_g), correlation length (ξ) and long period (L) estimated in the later stage of isothermal crystallization process for iPP sample (T_c = 127 °C).

Figure 2 shows the time dependence of infrared spectra measured for iPP sample in the isothermal crystallization process at T_c = 127 °C. The timing of detection of crystallization-sensitive band is different. For example, the band at 998 cm^{-1} started to increase in intensity at around 20 sec after the temperature reached T_c. The 841 cm^{-1} band was detected at ca. 30 sec. Another band at 1220 cm^{-1} appeared after ca. 80 seconds. These bands have different critical sequence length m of 10, 14, and longer than 15 monomeric units, respectively.

Figure 3 compares the time dependence of infrared and X-ray data in the isothermal crystallization process of iPP sample at T_c = 127 °C. After the sample temperature reached the T_c at 0 sec, the amorphous band decreased in intensity and the helical bands at 998, 841 and 1220 cm^{-1} started to appear around 20 sec, 30 sec and 80 sec, respectively. The 998 cm^{-1} band can be detected when the regular helical form is longer than the helix of 10 monomeric units (critical sequence length m). Similarly, the band at 841 cm^{-1} for m = 14 and the band at 1220 cm^{-1} for $m > 15$. The difference of timing of detection of these bands indicates that the short helices appear after the start of isothermal crystallization and they grow to longer helices with time. In Figure 3, the radius of gyration R_g of isolated domains decreases gradually in parallel to the appearance of these regular helices and their growth. These domains are correlated more strongly and the SAXS profile started to change at around 100 sec. The SAXS data were analyzed by Debye-Bueche equation. At around 100 sec, the ξ value was almost equal to R_g and it started to decrease linearly, during which the regular helical lengths continue to increase. The ξ value starts to deviate from the linear line at about 300 sec, where the peak corresponding to the lamellae started to appear as noticed from the rising up of the invariant Q the degree of crystallinity. At this point the WAXD peaks were also observed clearly. Figure 4 shows the schematic illustration of the structural changes in the isothermal crystallization process deduced from all the experimental data of FTIR, SAXS and WAXD.

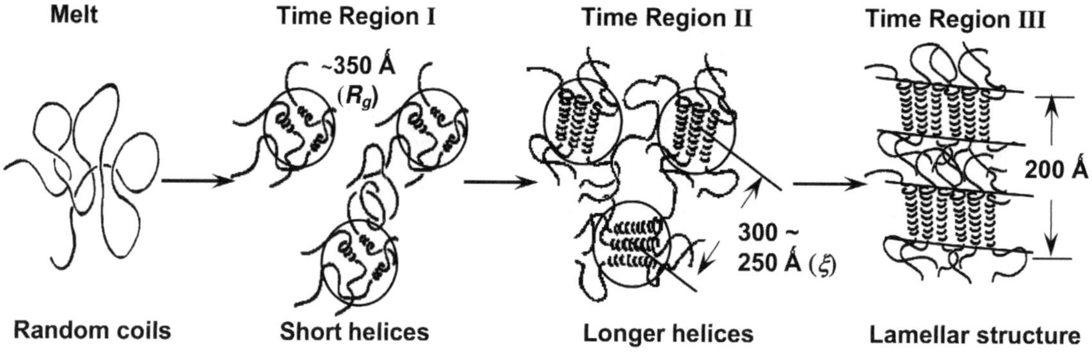

Figure 4. Schematic illustration of structural evolution process in iPP sample deduced from the combination of FTIR and SAXS/WAXD data (refer to Figure 3).

References

[1] A special issue on crystallization of polymers, *Faraday Discussions*, **1979**, *68*.
[2] Brückner, S.; Meille, S. V.; Petraccone, V.; Pirozzi, B. *Prog. Polym. Sci.* **1991**, *16*, 361.
[3] Heeley, E. L.; Maidens, A. V.; Olmsted, P. D.; Bras, W. *et al. Macromolecules* **2003**, *36*, 3656.
[4] Ryan, A. J.; Fairclough, J. P. A.; Terrill, N. J. *et al. Faraday Discussion* **1999**, *112*, 13.
[5] Zerbi, G.; Ciampelli, F.; Zamboni, V. *J. Polym. Sci.* **1963**, *C7*, 141.
[6] Kobayashi, M.; Akita, K.; Tadokoro, H. *Makromol. Chem.* **1968**, *118*, 324.
[7] Kissin, Y. V.; Rishina, L. A. *Eur. Polym. J.* **1976**, *12*, 757.

Structural Evolution in the Isothermal Crystallization Process of the Molten Nylon 10/10 Traced by Time-Resolved Infrared Spectral Measurements and Synchrotron SAXS/WAXD Measurements

Kohji Tashiro[1], Asami Nishiyama[1], Sawako Tsuji[1], Tomoko Hashida[1], Makoto Hanesaka[1], Shinichi Takeda[1], Cao Weiyu[1], Kummetha Raghunatha Reddy[1], Hiroyasu Masunaga[2], Sono Sasaki[2], Kazuki Ito[3], and Masaki Takata[2, 3]

[1]Department of Future Industry-oriented Basic Science and Materials, Toyota Technological Institute, Nagoya 468-8511, Japan.
[2]Japan Synchrotron Radiation Research Institute (JASRI), Kouto, Sayo-gun, Hyogo 679-5198, Japan.
[3]RIKEN Harima Institute/SPring-8, Kouto, Sayo-gun, Hyogo 679-5198, Japan
E-mail: ktashiro@toyota-ti.ac.jp

Abstract. The structural evolution in the isothermal crystallization process of nylon 10/10 from the melt has been clarified concretely on the basis of the time-resolved infrared spectral measurement as well as the synchrotron wide-angle and small-angle X-ray scattering measurements. Immediately after the temperature jump from the melt to the crystallization point, the isolated domains consisting of the hydrogen-bonded random coils were formed in the melt, as revealed by Guinier plot of SAXS data and the infrared spectral data. With the passage of time these domains approached each other with stronger correlation as analyzed by Debye-Bueche equation. These domains transformed finally to the stacked crystalline lamellae, in which the conformationally-regularized methylene segments of the CO sides were connected each other by stronger intermolecular hydrogen bonds to form the crystal lattice.

1. Introduction

Aliphatic nylon consists of the amide groups giving strong intermolecular hydrogen bonds and the methylene segments giving the flexibility to the skeletal chain. A sensitive balance between these two factors affects the hierarchical structure and physical properties of nylon. In other words, it is needed to clarify the formation process of the hierarchical structure to control the physical properties of this polymer by taking these two important factors into consideration. In a long history of nylon, however, there had been reported almost no paper describing the structural changes during the crystallization process from the melt. In the present study we have revealed successfully and in a concrete manner the structural evolution process of nylon 10/10, as one of the most typical aliphatic nylons, in the isothermal crystallization process from the melt by performing the time-resolved measurements of infrared spectra (FTIR) and synchrotron wide-angle X-ray diffraction (WAXD)/small-angle X-ray scattering (SAXS) patterns.

2. Experimental

The isothermal crystallization experiments have been performed by changing the temperature of the sample rapidly (ca. 1000°C/min) from above the melting point to the predetermined crystallization point. During this temperature jump, the time-resolved measurements of FTIR, WAXD and SAXS were carried out at a constant time interval (every 1 – 2 sec). The simultaneous measurements of WAXD and SAXS were performed using synchrotron X-ray source in the beam line BL40B2 of SPring-8, Japan. The samples used in the experiments were nylon 10/10 -[-(NH$(CH_2)_{10}$NHCO$(CH_2)_8$CO-]- and nylon 10/10-d_{16} -[-(NH$(CH_2)_{10}$NHCO$(CD_2)_8$CO-]-. The latter or partially-deuterated sample was useful especially in the infrared spectrosocopic experiment since the infrared bands of methylene segments of the CO side were distinguished from those of NH side. The transition behaviours of these two samples were essentially the same each other.

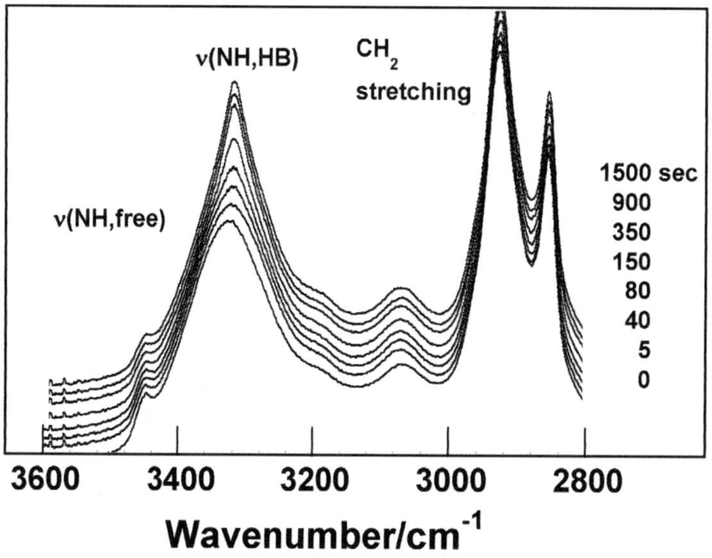

Figure 1. Time dependence of infrared spectra in the frequency region of 3600 – 2800 cm^{-1} measured for nylon 10/10-d_{16} in the isothermal crystallization at 176°C.

3. Results and Discussion

3.1 Time Dependence of Infrared Spectra

Figure 1 shows the time dependence of infrared spectra measured in the isothermal crystallization at 176°C. Figure 2 shows the time dependence of vibrational frequency and intensity of the infrared bands of NH stretching mode (3300 – 3440 cm^{-1}) and the CD_2 group mode (1065 cm^{-1}). In the melt before the temperature jump, the NH stretching band of the free hydrogen bonds [ν(NH, free)] was detected clearly at 3430 cm^{-1}. At the same time, the ν(NH) band corresponding to the hydrogen-bonded amide groups (3300 cm^{-1}) was also detected, though it was rather broad in the melt. In this way, the intermolecular hydrogen bonds are existent still in the melt. This is a key point to understand the crystallization behaviour of nylon 10/10 as will be discussed in a later section. These hydrogen

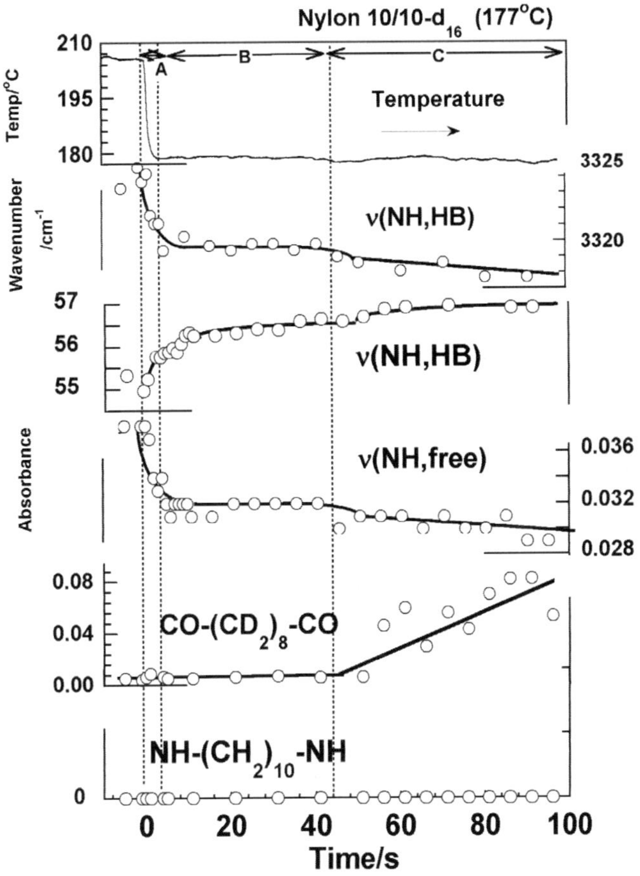

Figure 2. Time dependence of vibrational frequency and absorbance estimated for the various infrared bands of nylon 10/10-d_{16} in the isothermal crystallization at 176°C.

bonds became stronger and their relative amount increased drastically immediately after the temperature jump. On the way of crystallization in a time region from 10 to 60 sec the hydrogen bonds were stabilized for a while (a plateau region observed for NH stretching bands in Figure 2). After that they started to become stronger furthermore, just when the methylene segments of the CO side were regularized in conformation. The methylene segments of the NH side were still in the disordered state [1].

3.2 Time Dependence of WAXD and SAXS Profiles

Figure 3 shows the time dependence of WAXD and SAXS profile in the isothermal crystallization process of nylon 10/10 at 176°C. For the quantitative analysis of SAXS data in the time region of 0 – 60 sec, we assumed that (i) in the earliest stage the locally-formed weak hydrogen bonds in the melt might cause the generation of some domains with relatively high density. These domains are isolated each other. (ii) As the time passes, these domains may transform to the stacked lamellae with relatively strong correlation. Then we analyzed the SAXS data on the basis of Guinier (for isolated domains; scattering vector $q < 0.007$ Å^{-1}) and Debye-Bueche theories (for correlated domains; $q > 0.007$ Å^{-1}) as well as the correlation function concerning the stacked lamellar structure.

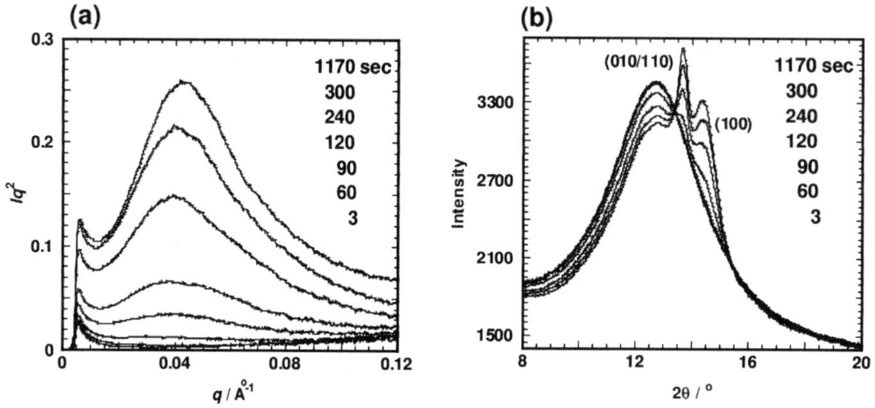

Figure 3. Time dependences of (a) SAXS and (b) WAXD profiles of nylon 10/10 measured in the isothermal crystallization process at 176°C.

As shown in Figure 4 (a), the Guinier plot [ln(I) vs q^2] gave the size of domain (or radius of gyration R_g) consisting of relatively strongly hydrogen-bonded coils [2]. As shown in Figure 4 (b) the Debye-Bueche plot [$I^{-1/2}$ vs q^2] gave the averaged distance ξ (correlation length) between these domains [3]. Figure 5 shows the time dependence of the thus estimated R_g and ξ values. Figure 6 illustrates the relation between R_g, ξ (and long period) given in Figure 5. As mentioned above, the R_g is the size of domain with almost no correlation with the neighboring domains. Once the crystallization is started, these domains approach more closely to each other due to the thermal contraction of the whole system, and at the same time the inner structure becomes more regular due to the increasing strength of hydrogen bonds, as revealed by the infrared spectral data. Once the correlation length, a measure of distance between the correlated neighboring domains, is close to the R_g [see Figure 6 (b)], the correlation between these regular parts start to increase. The correlation becomes stronger and approaches finally to the long period of the stacked lamellae. (It is noticed that the R_g was almost constant during this process as long as the Guinier plot was assumed to be applicable even for such a

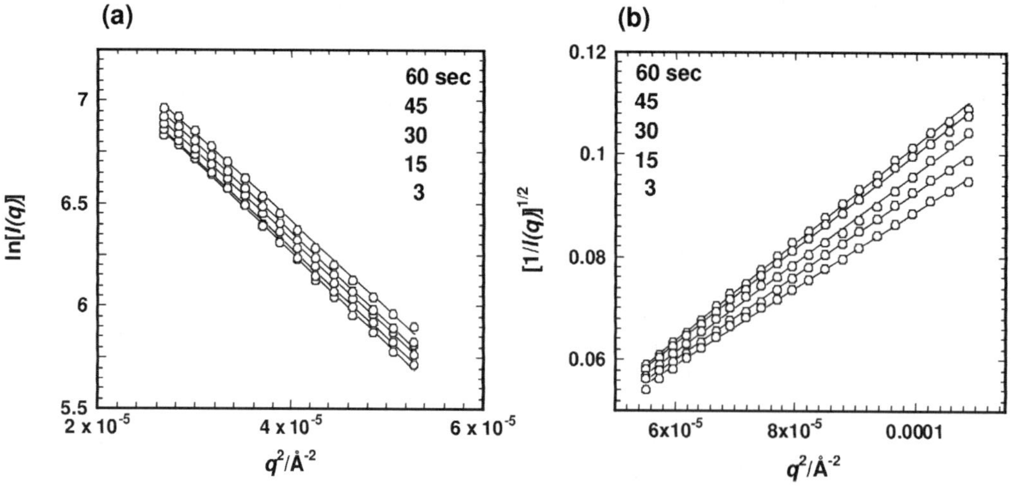

Figure 4. (a) Guinier and (b) Debye-Bueche plots made for SAXS data of nylon 10/10 obtained in the isothermal crystallization at 176°C.

weakly correlated domain structure as an approximation.) As the time passed beyond 60 sec, the ξ value started to deviate from the linear straight line and transformed to the long period of stacked lamellae, about 240 Å. The SAXS data in this time region was analyzed by calculating the correlation function for the stacked lamellar structure [4], and the invariant Q was estimated as seen in Figure 5.

3.3 Structural Evolution Process of Nylon 10/10

From Figure 5, we may draw the structural evolution process of nylon 10/10 in the isothermal crystallization from the melt as illustrated in Figure 7. At the earliest stage of crystallization from the melt, the intermolecular hydrogen bonds become stronger steeply and the amide groups of free hydrogen bonds decrease in population. These hydrogen bonds induce the local creation of isolated domains of relatively high density. As the time passes furthermore, these domains approach each other because the sample contracts gradually in the solidification process, during which the intermolecular hydrogen bonds are stabilized temporarily (10 – 60 sec region). After that the domains gather together more closely to form the stacked lamellar structure, in which the conformationally regularized methylene segments (the methylene segments of the CO side, strictly speaking) are packed in parallel and linked together by the strong intermolecular hydrogen bonds to form the crystal lattices. (According to the infrared spectra measured for the partially-deuterated nylon 10/10 sample, the methylene segments of the NH side are still in the conformationally-disordered state even when the methylene segments of the CO side are already regularized. The reason for the difference in conformational regularization between these two kinds of methylene segments is not solved yet [1].)

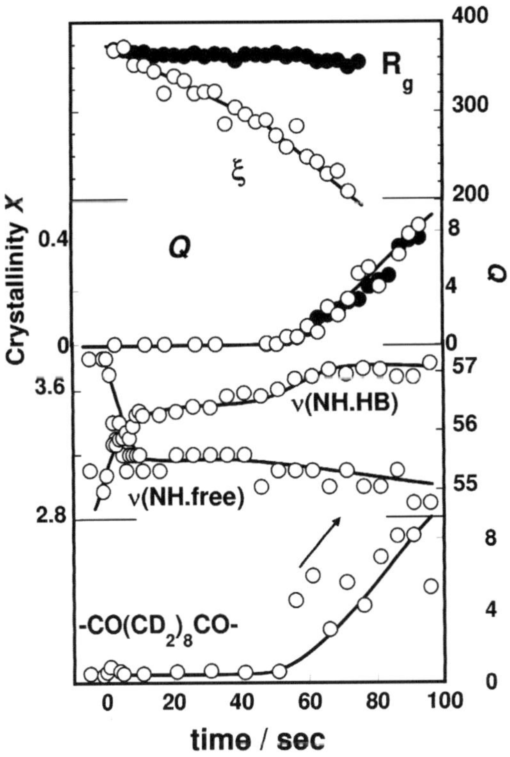

Figure 5. Time dependence of R_g and ξ compared with the infrared band data taken for nylon 10/10 in the isothermal crystallization at 176°C.

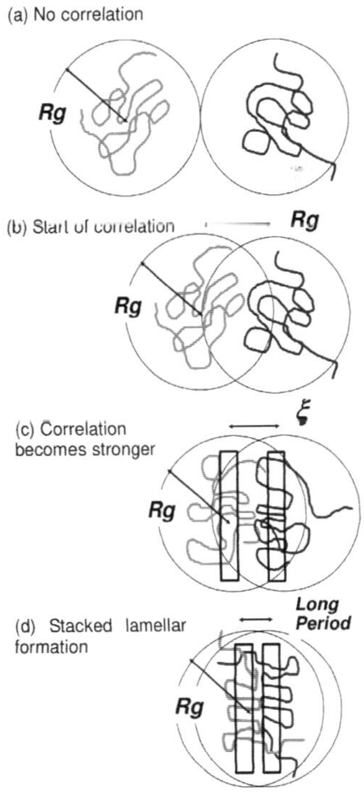

Figure 6. Schematic illustration of relationship between R_g, ξ and long period.

This structural evolution process is appreciably different from that detected for polyolefin like polyethylene, for example. In the latter case the random coils in the melt started to regularize after the temperature jump, and the trans-rich zigzag-like chains are formed with some conformational disorders. These chains gather together to form the pseudo-hexagonal lattice, which transforms gradually to the orthorhombic cell consisting of regular all-trans zigzag chain stems [5]. In this case the molecular chains are connected with weak van der Waals interactions. In the aliphatic nylon case, the methylene segments are constrained more or less due to the intermolecular hydrogen bonds even in the molten state. Therefore, some domains of comparatively high density are existent from the starting point of crystallization, making it easier to form the aggregation state of conformationally-regularized stems linked side by side with the strong intermolecular hydrogen bonds.

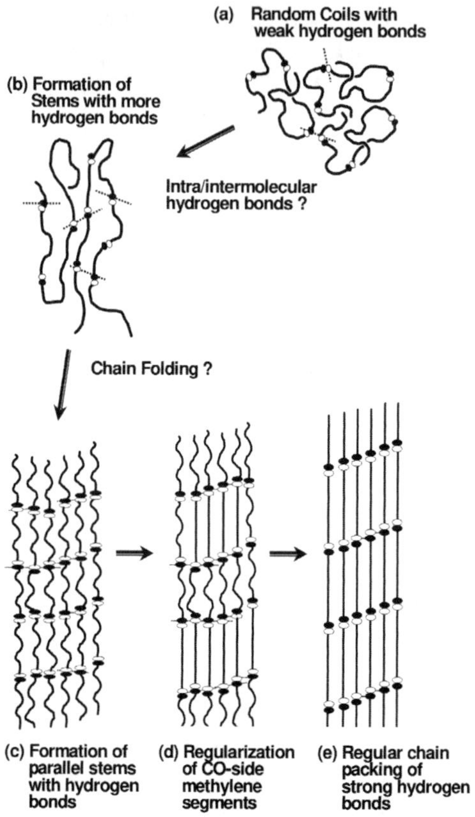

Figure 7. Illustration of structural evolution of nylon 10/10 in the isothermal crystallization.

4. Conclusions

In the present report the structural evolution process of nylon 10/10 in the isothermal crystallization phenomenon has been revealed by measuring the time dependences of infrared spectra and synchrotron WAXD/SAXS data. The hydrogen bonds between the amide groups are existent even in the molted state although they are rather weak, which play an important role in the formation of domains with higher density. After the passage of some time, these domains gather together and transform to the stacked lamellar structure with regularly packed zigzag methylene segments combined with the strong intermolecular hydrogen bonds. This crystallization behavior is quite contrast to the case of polyethylene without any strong intermolecular hydrogen bonds. The difference in processability between these two kinds of polymer might reflect such a difference in crystallization process from the molten state.

Acknowledgements

This study was financially supported by a MEXT "Collaboration with Local Communities" project (2005 – 2009).

References

[1] Tashiro, K.; Yoshioka, Y. Polymer **2004**, 45, 6349
[2] Guinier, A.; Fournet, G. *Small angle X-ray Scattering*, John Wiley & Sons, Inc.: London, **1955**.
[3] Debye, P.; Bueche, A. M. *J. Appl. Phys.* **1949**, 20, 518
[4] Strobl, G. R.; Schneider, M. *J. Polym. Sci., Polym. Ed.*, **1980**, 18, 1343
[5] Sasaki, S.; Tashiro, K.; Kobayashi, M.; Izumi, Y.; Kobayashi, K. *Polymer* **1999**, 40, 7125

Journal of Physics: Conference Series **184** (2009) 012003 doi:10.1088/1742-6596/184/1/012003

First Detection of Lamella-Gyroid-Cylinder Phase Transition of Neat Polyethylene-Poly(ethylene Oxide) Diblock Copolymers on the Basis of Synchrotron WAXD/SAXS and Infrared/Raman Spectral Measurements

Cao Weiyu[1], Kohji Tashiro[1], Makoto Hanesaka[1], Shinichi Takeda[1], Hiroyasu Masunaga[2], Sono Sasaki[2] and Masaki Takata[2]

[1] Department of Future Industry-oriented Basic Science and Materials, Toyota Technological Institute, Nagoya 468-8511, Japan
[2] Japan Synchrotron Radiation Research Institute (JASRI), Kouto, Sayo-gun, Hyogo, 679-5198, Japan
E-mail: ktashiro@toyota-ti.ac.jp

Abstract. The phase transition behaviour of polyethylene-*b*-poly(ethylene oxide) (PE-*b*-PEO) diblock copolymer with relatively short chain lengths has been studied on the basis of temperature dependent infrared and Raman spectral measurements and synchrotron WAXD/SAXS simultaneous measurements, from which the concrete structural changes were deduced successfully from the various levels of molecular chain conformation, chain packing mode and higher-order structure. The higher-order structure has been found to transform between lamella, perforated lamella, gyroid, cylinder and sphere structures. The inner structural changes occurring in the polyethylene and poly(ethylene oxide) parts have been related with these morphological changes. The morphological transition from lamella to gyroid occurs with keeping the crystalline state of polyethylene parts. This apparently curious transition can be interpreted reasonably by assuming the thermally-activated chain motion in the crystal lattice, which may play an important role as a trigger to induce the morphological change from lamella to gyroid. This idea was supported by the measurement of half-width of Raman anti-symmetric CH_2 stretching band sensitive to the thermal mobility of alkyl chains.

1. Introduction

In a diblock copolymer consisting of two immiscible polymer components, the phase separation on the macroscopic level is prevented by a direct covalent linkage of these polymer chains. The tendency of segregation of these chain blocks leads to a so-called microphase separation, where the separation of the components occurs in nanoscale domains[1]. As a result, they can exhibit various morphologies such as lamella, gyroid, cylinder, sphere structures. The two polymer components studied for a long time were mainly amorphous. In these years the crystallline-amorphous and crystalline-crystalline diblock copolymers have been investigated actively because they exhibit the various types of phase transitions in the crystalline regions which should be related with the drastic morphological changes. But the details of transition behaviors of crystalline-crystalline diblock copolymers have not yet been clarified enough satisfactorily.

Polyethylene-*block*-poly(ethylene oxide) is one of the typical crystalline-crystalline diblock copolymers. In the aqueous solution it exhibits the various kinds of morphology like cylinder, gyroid

Journal of Physics: Conference Series **184** (2009) 012003
doi:10.1088/1742-6596/184/1/012003

and lamella depending on the concentration and temperature[2]. About the neat PE-*b*-PEO system without any water content, Sun et al.[3] studied the low-molecular-weight PE-*b*-PEO diblock copolymer and proposed the lamellar structure in a wide temperature region without any transition to the other types of morphology. Castillo et al.[4] reported a possibility of sphere phase in a high temperature region, but the SAXS profile was too broad to identify the detailed morphology because of relatively high content of side branches. We have investigated the phase transition behaviors of PE-*b*-PEO diblock copolymers with different chain length combinations on the basis of simultaneous synchrotron WAXD/SAXS measurements as well as vibrational spectral data[5]. We observed successfully the transitions between lamella, gyroid, cylinder and sphere morphologies for PE-*b*-PEO copolymers with the segmental lengths of PE and PEO similar to those utilized by the previous researchers.[3] The results obtained for PE-*b*-PEO diblock copolymer with PE and PEO segmental lengths of 39 and 86 monomeric units (E39EO86), respectively, will be mainly described here.

2. Experimental

The sample E39EO86 was purchased from Polymer Source Inc. The molecular weight distribution was 1.10. The infrared spectral measurement was performed using a Varian FTS7000 Fourier-transform infrared spectrometer at a resolution power 1 cm^{-1}. The simultaneous measurement of 2D-WAXD and SAXS data was performed at BL40B2 beamline at SPring-8 (Japan Synchrotron Radiation Research Institute, Japan). The thermal behavior was investigated using a TA Instruments DSC Q1000 under nitrogen gas atmosphere at a heating rate of 10°C/min.

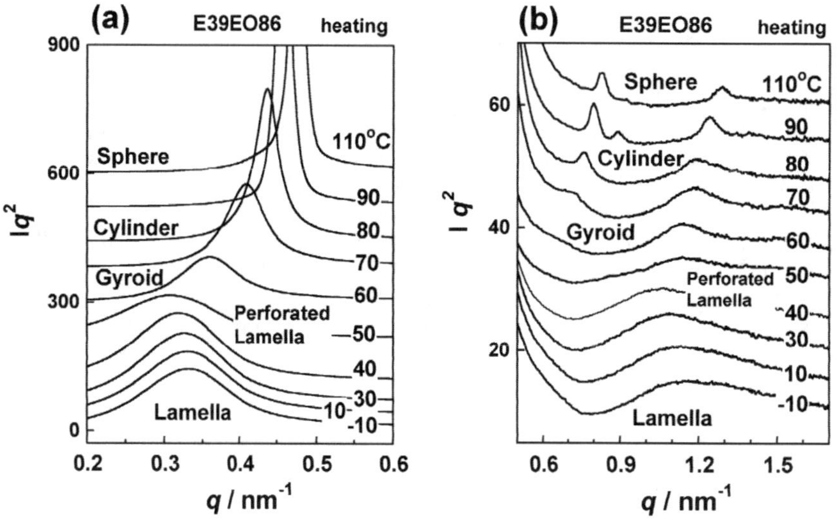

Figure 1. Temperature dependence of SAXS profile of E39EO86 measured in the heating process: (a) q=0.2-0.6nm^{-1} and (b) 0.5-1.7nm^{-1}.

3. Results and Discussion

3.1. Morphological changes

The SAXS profiles of E39EO86 obtained in the heating process are shown in Figure 1. At room temperature, the peaks were detected at the positions of the ratio 1 : 2 : 3, indicating a lamellar lamellar morphology. By heating the sample to above 60°C, the ratio of the peak positions changed to

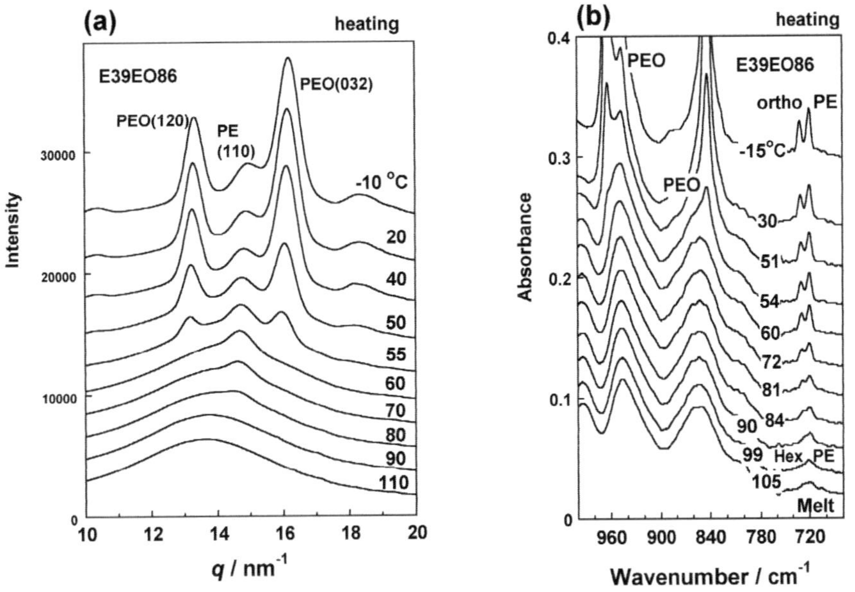

Figure 2. Temperature dependence of (a) WAXD profile and (b) IR spectra in the frequency region of 1000~680cm^{-1} of E39EO86 measured in heating process.

1 : 1.15 : 1.53 : 1.63 : 2, which corresponds to the gyroid structure[1]. Before transforming to gyroid structure, the SAXS peaks characteristic of a metastable perforated lamella was detected although the details were not clear enough well. By heating furthermore, the hexagonally-packed cylindrical phase was observed at around 90°C, giving the ratios of peak positions 1 : 1.73 : 2 : 2.65 : 3. As the temperature was increased to above 100°C, the sphere structure started to appear. In the cooling process these morphological changes were detected almost reversibly with some thermal hysteresis.

3.2 Structural Changes in Crystal Lattice

Figure 2 (a) and (b) show the temperature dependences of WAXD profile and infrared spectral pattern measured in the heating process. The X-ray reflection peaks intrinsic to PEO crystal disappeared above 55°C. The reflections of PE orthorhombic phase were observed up to 80°C. The infrared spectra detected not only the melting of PEO crystal at around 54°C but also the transition of PE segments from the orthorhombic crystal to the pseudo-hexagonal or rotator phase[8] as seen in the disappearance of a pair of bands at 730 and 720 cm^{-1} and the appearance of a single band at 721 cm^{-1} at around 90°C. Figure 3 shows the comparison of DSC, IR, WAXD and SAXS data as a function of temperature. Once the PEO segments are melted, the transition from lamella to gyroid occurs. It should be noticed here that the PE segments are still in the orthorhombic crystalline phase. Similarly, the gyroid to cylinder transition occurs while the orthorhombic PE crystal transfers to the pseudo-hexagonal form. In this way, the morphological transition is found to occur even in the solid state of PE segmental parts. It may be speculated that the PE chains are thermally activated and can move along the interfacial boundary surface to cause the lamella-to-gyroid transition. This hypothesis could be confirmed using the half-width of Raman band at 2880 cm^{-1} [the methylene anti-symmetric stretching mode $\nu_{as}(CH_2)$] which is sensitive to the thermal mobility of methylene chains[6]. The Raman spectra were measured for the PE-*b*-deuterated PEO block copolymer E39dEO25 with the same PE segmental length as that of E39EO86. The deuterated sample was needed to separate the $\nu_{as}(CH_2)$ band of PE from those of PEO species by deuteration of ethylene oxide parts. Figure 4 shows the experimental data collected for E39dEO25 sample. The half-width of $\nu_{as}(CH_2)$ band was

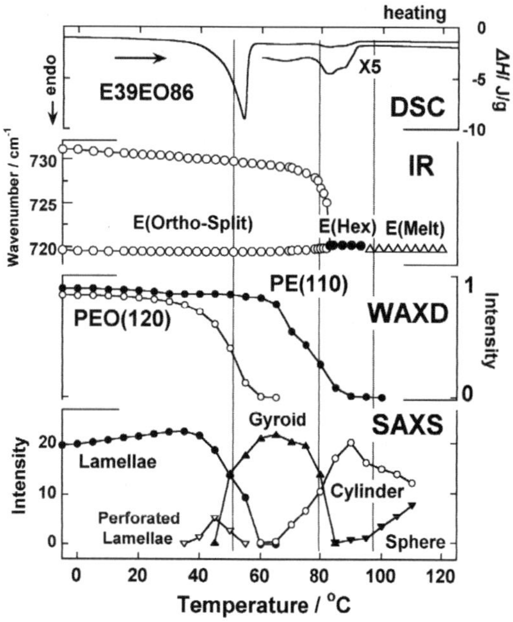

Figure 3 Comparison of the DSC, IR, WAXD and SAXS data collected for E39EO86 in the heating process.

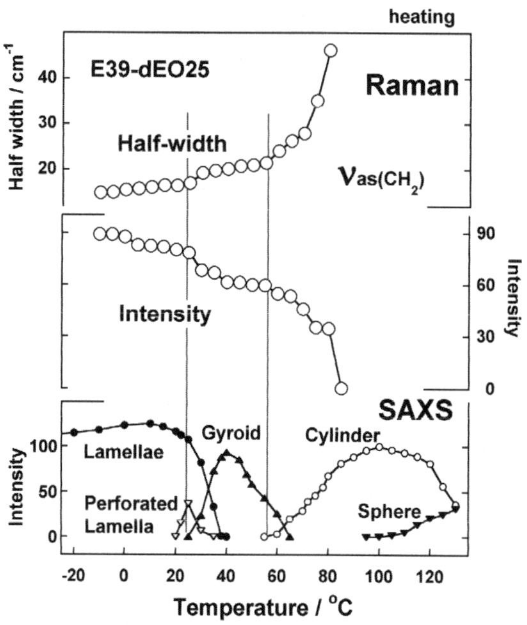

Figure 4 Comparison of the half-width and integrated intensity of Raman $\nu_{as}(CH_2)$ band and SAXS intensity estimated for E39-dEO25 in the heating process.

found to increase largely just when the lamella-to-gyroid and gyroid-to-cylinder phase transitions occurred. That is to say, the enhancement of thermal mobility occurred in the orthorhombic crystal region of PE segmental parts, resulting in the transition of lamella-to-gyroid phases.

4. Conclusions

As reported in the present paper, the complicated morphological changes were found out for PE-*b*-PEO diblock copolymer with relatively short segmental lengths. By viewing from the various levels these morphological transitions were related intimately with the structural changes occurring in each chain parts. The lamella-gyroid (and gyroid-cylinder) transition was observed even when PE parts were still in the crystalline state, unexpected from the reports on many crystalline-crystalline diblock copolymers with lamellar morphology only. This phenomenon was ascribed to the activation of thermal motion of PE chains in the crystalline state.

References

[1] Hamley, I. W. *The Physics of Block Copolymer*; Oxford, University Press: New York,**1998**

[2] Rancon, Y.; Charvolin, J. *J. Phys. Chem.* **1998**, 92, 2646

[3] Sun, L.; Liu, Y.; Zhu, L.; Hsiao, B. S.; Avila-Orta, C. A. *Polymer* **2004**, 45, 8181; *Macromol. Rapid Commun.* **2004**, 25,853

[4] Castillo, R.V.; Arnal, M. L.; Muller, A. J.; Hamley, I.W.; Castelletto, V.; Schmalz, H.; Abetz, V. *Macromolecules* **2008**, 41, 879

[5] Cao, W-Y.; Tashiro, K.; Hanesaka, M.; Takeda, S.; Masunaga, H.; Sasaki, S.; Takata, M. *J. Phys. Chem.* **2009**, in press

[6] Snyder, R. G.; Scherer, J. R.; Gaber, B. P. *Biochim. Biophys. Acta* **1980**,601, 47

Journal of Physics: Conference Series **184** (2009) 012004 doi:10.1088/1742-6596/184/1/012004

Distribution of Glass Transition Temperature T_g in a Polymer Thin Film by Neutron Reflectivity

K Kawashima[1], R Inoue[1,2], T Kanaya[1,4], G Matsuba[1], K Nishida[1], M Hino[3]

[1] Institute for Chemical Research, Kyoto University, Uji, Kyoto-fu 611-0011, Japan
[2] Institut für Festkörerforschung, Forschungszentrum Jülich, D-52425 Jülich, Germany
[3] Research Reactor Institute, Kyoto University, Kumatori-cho, Sennan-gun, Osaka-fu 590-0494, Japan

E-mail: kanaya@scl.kyoto-u.ac.jp

Abstract. We report neutron reflectivity results on a three-layer polystyrene thin film on Si wafer, consisting of alternative stacking of deuterated layer ~20 nm thick and hydrogenated layer ~30 nm thick. In the experiments we have evaluated the film thickness of each layer and the surface roughness and interfacial widths as a function of temperature below and above the glass transition temperature T_g. It was found that the top layer has T_g lower than the bulk T_g by ~18 K, and the middle layer has almost the same T_g as the bulk. On the other hand, the bottom layer did show very high T_g above 130 °C. The results show the distribution of T_g in the thin film. It was also found that the interfacial width between the top and middle layers increased more rapidly than that between the middle and bottom layers above ~110 °C, showing higher mobility of polystyrene in the top layer because of the lower T_g.

1. Introduction

Recent extensive studies [1] on polymer thin films have revealed very interesting but unusual properties of polymer thin films. Glass transition of thin films is one of the most interesting topics in the field because many properties such as mechanical and thermal properties change drastically at the glass transition temperature T_g. In a pioneering work by Keddie et al. [2] it was reported that the glass transition temperature of polystyrene thin films supported on silicon substrate decreases with decreasing thickness below 40nm. Similar results were reported by many researchers on T_g of polystyrene thin films supported on silicon substrate [1]. On the other hand, T_g increases with decreasing the film thickness for poly(methyl methacrylate)(PMMA) supported on silicon substrate [3], suggesting strong interactions between them. Thus, surface and/or interfacial effects seem to play an important role on glass transition.

In order to reveal the cause of the decrease in T_g Kajiyama et al. [4-6] have studied T_g of the surface layer of polystyrene thin films to find a lower glass transition temperature T_g, and hence as the film thickness decreases the fraction of the surface layer increases, resulting in the reduction in T_g. This is very normal explanation for the reduction in T_g for thin films. However, Forrest et al. [7] carried out ellipsometry measurements on polystyrene thin films to evaluate the glass transition temperature T_g as well as the time constant of interface healing in bi-layer polystyrene films to find that as the film

[4] To whom any correspondence should be addressed.

thickness decreases the glass transition temperature T_g decreases while at the same time the time constant for the interface healing process increases. Based on the observations they pointed out that it was not always possible to relate the lower T_g to the higher mobility. Similar results can be found in inelastic neutron scattering measurements on polymer thin films. Soles et al. [8,9] have measured the mean square displacement $\langle u^2 \rangle$ of polymers, which is a dynamic property of polymers, and found that the mean square displacement $\langle u^2 \rangle$ decreased with the film thickness while the glass transition temperature estimated by X-ray reflectivity decreased. These observations again suggest that the lower glass transition T_g does not always mean the higher mobility. It should be noted that polymers have wide distribution of the relaxation time in time scale as well as in length scale. Therefore, in order to solve the contradiction mentioned above we have to reveal the distribution of T_g as well as the dynamic nature. In other words, revealing heterogeneous structure and/or dynamics in polymer thin films is essential to understand the unusual properties. This has been already pointed out by de Gennes [10], and he mentioned that further experiments should not aim at the determination of a single T_g, but at a distribution of T_g's. Recent woks by Torkelson et al. [11] along the direction are remarkable. Using a fluorescence technique they examined glass transition temperature at surface, interface and in between of multilayer films.

In this report we have studied the distribution of glass transition temperature T_g in a three-layer polystyrene thin film by neutron reflectivity, aiming to elucidate the heterogeneous structure or multi-layer structure of polymer thin films.

2. Experimental

For neutron reflectivity (NR) measurements we used deuterated polystyrene (d-PS) and hydrogenated polystyrene (h-PS) with molecular weight $M_w = 7.31 \times 10^5$ and $M_w = 7.69 \times 10^5$, and molecular weight distributions were $M_w/M_n = 1.18$ and 1.08, respectively. Here, M_w and M_n are the weight average and number average molecular weights, respectively. Unperturbed radii of gyration R_g are 23.3 and 23.9 nm for d-PS and h-PS, respectively. The bulk glass transition temperatures T_g of h-PS and d-PS are 100°C and 103 °C, respectively, which were determined by DSC measurements.

The three-layer thin film was prepared as follows. First, a bottom d-PS film was prepared on Si (111) surface by spin-coating the toluene solution at 2000 rpm. The second h-PS thin film was prepared on another Si surface and then removed onto water surface. The floated h-PS thin film on water surface was collected on the first d-PS surface. The third d-PS film was prepared on the second h-PS film in the same way as for the second film. The three-layer film thus prepared was annealed carefully. First it was annealed at 90 °C which was ~10 °C below the bulk T_g after drying in vacuum at room temperature for two days, and then heated up to 130 °C at a hearing rate of 10 °C/min, and after keeping the sample at 130 °C for 5 min it was cooled down below T_g at a cooling rate of -10 °C/min. Thus, we did very rapid annealing to minimize the inter-diffusion between d-PS and h-PS.

NR measurements were performed in a temperature range from 25 to 130 °C, covering the bulk glass transition, using MINE spectrometer at JRR-3 reactor in Tokai. The incident wavelength of neutron was 0.8 nm and the range of length of scattering vector q was 0.08 to 0.7 nm^{-1} in the experiments. It took about one hour to measure one reflectivity profile at a given temperature.

3. Results and Discussion

In this work we performed neutron reflectivity measurements on a three-layer thin film, consisting of alternative stacking of d-PS layer ~20 nm thick and h-PS layer ~30 nm thick. The total thickness is ~65 nm. Deuterated and hydrogenated PS's are almost identical in their properties, but the scattering cross-sections for neutrons are so different that neutrons can distinguish the deuterated and hydrogenated PS layers easily. Figure 1 shows neutron reflectivity profiles from the three-layer thin film at 30, 60, 85 and 110 °C. It is very clear that the reflectivity profiles gradually shifts to lower q with increasing the temperature, showing thermal expansion of the film. The observed reflectivity profiles were analyzed by a program Motofit developed based on Parratt recursion method [12]. The results of fits are shown by solid curves in Figure 1. The fitness is good. We have evaluated thickness of each layer, and surface roughness and interfacial width as a function of temperature.

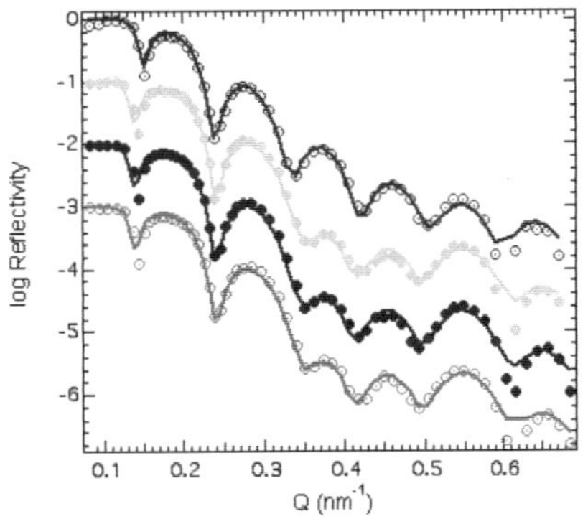

Figure 1. Neutron reflectivity profiles of three-layer polystyrene thin film at 30, 60, 85 and 110 °C from bottom to top. Solid curves are the results of fits.

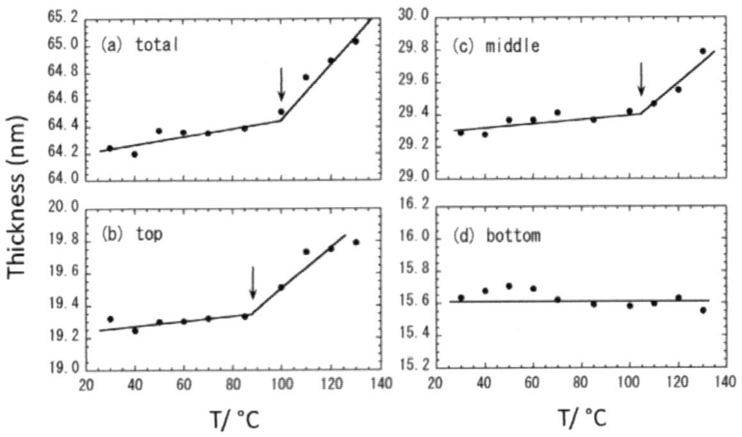

Figure 2. Temperature dependence of film thickness. (a) total, (b) top layer, (c) middle layer, (d) bottom layer.

The temperature dependence of the total film thickness is shown in Figure 2(a). The total film thickness increases with temperature in the glassy state up to ~100 °C and begins to increase more steeply above ~100 °C, showing the glass transition temperature T_g is ~100 °C which is almost the same as the bulk glass transition temperature (= 103 °C). It is noted that in this thin film we did not observe the decrease in thickness slightly below T_g, which was typically observed in not-well annealed thin films [13] and assigned to structure relaxation due to lack of annealing. The present observation suggests that the annealing condition in this experiment is enough to avoid the structure relaxation near T_g. The temperature dependences of the thickness for the top (surface), the middle and the bottom layers are shown in Figures 2(b), (c) and (d), respectively. The surface layer increases in thickness with the thermal expansivity of ~1.3 x 10^{-4} K^{-1}, which is close to the expected value (= 1.1 x 10^{-4} K^{-1} [14]), while it begins to increase above ~85 K more steeply, showing that the glass transition temperature of the surface layer is ~18 K lower than the bulk T_g. The middle layer shows a very similar thermal expnsivity in the glassy state with the top layer, but the glass transition temperature is ~105 K which is very close to or little bit higher than the bulk T_g. The middle layer behaves as the bulk layer. On the other hand, the bottom layer shows very different temperature dependence of the thickness from other layers. The film thickness is almost independent of temperature below ~130 °C.

In the temperature range examined we did not observe the increase in thickness, suggesting that the glass transition temperature of the bottom layer must be higher than 130 °C. The present results clearly show that the thin film has a distribution of T_g even though the T_g of the total thickness is the same as the bulk T_g.

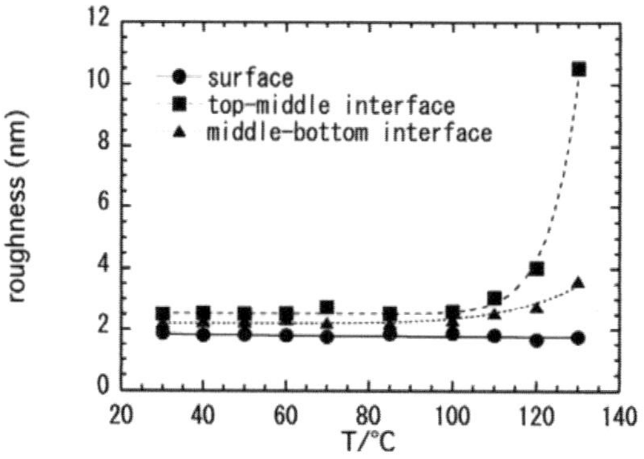

Figure 3. Temperature dependence of surface roughness, and top-middle and middle-bottom interfacial width.

Figure 3 shows the temperature dependence of surface roughness and interfacial widths of the three-layer film. The surface roughness of the top layer is almost independent of temperature. On the other, the interfacial width between the top and the middle layers (top-middle interfacial width) and that between the middle and the bottom layers (middle-bottom interfacial width) increase with temperature, especially above the glass transition temperature. These increases are caused by the inter-diffusion of d-PS and h-PS. As the temperature exceeds ~100 °C, the top-middle interfacial width increases much faster than the middle-bottom interfacial width with temperature, suggesting that the PS diffusion at the top-middle interface is faster than at the middle-bottom interface. This must be due to the lower T_g of the top layer, and the slow diffusion at the middle-bottom interface is due to the higher T_g of the bottom layer. Note that the top-middle interfacial width at 130 °C is about 10 nm. This value is almost the limit to distinguish the top layer and the middle layer. To study temperature dependence of thickness for each layer at higher temperature above ~130 °C faster measurements are absolutely necessary.

References

[1] Jones R L and Richards R W 1999 *Polymers at Surface and Interfaces* (Cambridge: Cambridge University Press)
[2] Keddie J L, Jones R A and Cory R A 1994 *Europhys. Lett.* **27** 59
[3] Keddie J L, Jones R A and Cory R A 1994 *Faraday Discuss.* **98** 219
[4] Kajiyama T, Tanaka K, Satomi N and Takahara A 1998 *Macromolecules* **31** 5150
[5] Kajiyama T, Tanaka K and Takahara A 1998 *Polymer* **39** 4665
[6] Satomi N, Takahara A and Kajiyama T 1999 *Macromolecules* **32** 4474
[7] Fakhraai Z, Valadkhan S and Forrest J A 2005 Eur. *Phys. J. E* **18** 143
[8] Soles C L, Douglas J F and Wu W –l 2002 *Phys. Rev. Lett.* **88** 37401
[9] Soles C L Douglas J F, Wu W -l, Peng H and Gidley D W 2004 *Macromolecules* **37** 2890
[10] de Gennes P G 2000 Eur. *Phys. J. E* **2** 201
[11] Priestley R D, Ellison C J, Broadbelt L J and Torkelson J M 2005 *Science* **309** 456
[12] Parratt L G 1954 *Phys. Rev.* **95** 359
[13] Miyazaki T, Nishida K and Kanaya T 2004 *Phys. Rev. E* **69** 022801
[14] Miyazaki T, Nishida K and Kanaya T 2004 *Phys. Rev. E* **69** 061803

Journal of Physics: Conference Series **184** (2009) 012005
doi:10.1088/1742-6596/184/1/012005

Effect of Reflected Waves on the GISAXS Analysis of As-grown Capped Ge Nanodots

H. Okuda[1*], K. Kuno[1], S. Ochiai[1], N. Usami[2], K. Nakajima[2], O. Sakata[3], S. Sasaki[3] and M. Takata[3]

[1]Kyoto University, Kyoto 606-8501 Japan, [2]IMR, Tohoku University, 980-8577 Sendai, Japan. [3]JASRI Sayo, Hyogo 679-5198 Japan.

Email: okuda@materials.mbox.media.kyoto-u.ac.jp

Abstract. Effect of layer structures on the GISAXS intensity analysis of Ge nanodots capped with Si layer has been examined by Born Approximation and Distorted Wave Born Approximation simulations. It is concluded that when the dots are small enough in the out-of-plane directions, most of the structural information are deduced by simple BA. The effect of refracted wave became important when the area of interest is close to the Yoneda line or the analysis deals with small difference in the intermediate q range.

1. Introduction

Grazing incidence small-angle X-ray scattering (GISAXS) has been used to examine the microstructures at the surface region. For the interpretation of the scattering patterns obtained for such surface nanostructures, the dynamical effect [1] [2] has been taken into account as an interesting phenomenon characteristic for GISAXS [3]-[5]. Owing to such developments, it has become a common understanding that a simulation taking into account the dynamical effect is indispensable in the analysis of GISAXS profiles. By publications by Lazzari et al. [6] [7], many structures listed in them are readily available for simulation. On the other hand, it is also clear that how the dynamical effect should be taken into account strongly depends on the nanostructure of individual sample and the detail of the analysis. In the early stage of GISAXS studies, researchers who are more interested in microstructure itself intentionally avoid the condition where dynamical effect is dominant [8]. We have reported[9],[10] that even a simple Born Approximation is sufficient for a simple size and shape analysis of capped nanodot if the height of nanodots are reasonably small. In the present study, effect of reflected wave on the GISAXS profile were examined for as-grown Ge nanodot capped with a thin Si layer.

2. Experimental and simulation procedures

2.1. Sample

The sample examined in the present work contains capped Ge nanodots having no apparent facets. Germanium nanodots were grown on a Si (001) substrate with a gas-source MBE at 783 K, and then capped with a Si layer with a thickness of about 40 nm. Samples used in the present measurements were cut from center part of 2 inch wafers, where uniformity of thickness was confirmed for the growth condition.

Journal of Physics: Conference Series **184** (2009) 012005 doi:10.1088/1742-6596/184/1/012005

2.2. Reflectivity/GISAXS measurements

Specular reflectivity was measured at Beam-line 13XU of SPring8, Japan. GISAXS measurements for the corresponding sample were made at the same beamline using a compact GISAXS apparatus on a multiaxis goniometer and also at BL15A of Photon Factory, Tsukuba, Japan using a conventional SAXS camera in a grazing-incidence geometry [9]. To examine in-plane anisotropy, several directions of incidence were chosen for GISAXS pattern measurements [10].

2.3. Simulations

Roughness and thickness of each layer in the sample were evaluated by using a least-square fit of the reflectivity curve. Then these parameters were used to calculate GISAXS pattern within a framework of DWBA. Since the scattering objects, Ge nanodots, are on a single plane, we assumed here that the scattered waves and these with reflections were coherent.

3. Results

Figure 1 gives a measured and a simulated specular reflectivity of the as-grown sample. The reflectivity curve shows two distinct characters, i.e., a short period oscillation and a slow modulation of the envelope. These two periods correspond to the total thickness of cap and dot layers, and the thickness of dot layer [11]. From a least-square fitting of the intensity, a simulation result suggests that the thickness of the cap Si layer and effective thickness of dot layers are 39.8 nm and 1.98 nm respectively. The roughnesses of the surface and cap/dot interface are estimated to be 0.5 nm and 0.9 nm respectively. These results suggest that the present sample has relatively smooth surface, although the roughness is not as good as a well-defined clean Si surface.

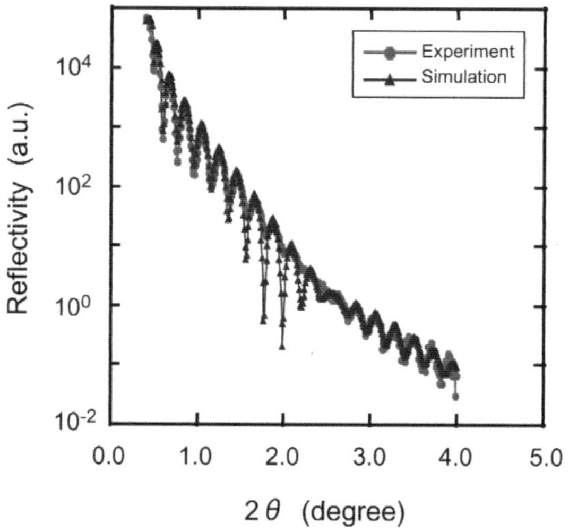

Figure 1. Specular reflectivity of the capped Ge nanodots measured at BL13XU of SPring8. Triangular points correspond to a simulation results obtained by least-square fitting.

A GISAXS pattern obtained for in-plane angle of 45 degrees between the incident beam and [110] direction of the sample, ϕ, is shown in Fig. 3. There are two main origins of diffuse scattering in the sample. On is the Ge nanodots which are of our main interests, and the other is surface and interface diffuse scattering. The latter has intensively analysed in the multilayer structures in the last two decades. From the

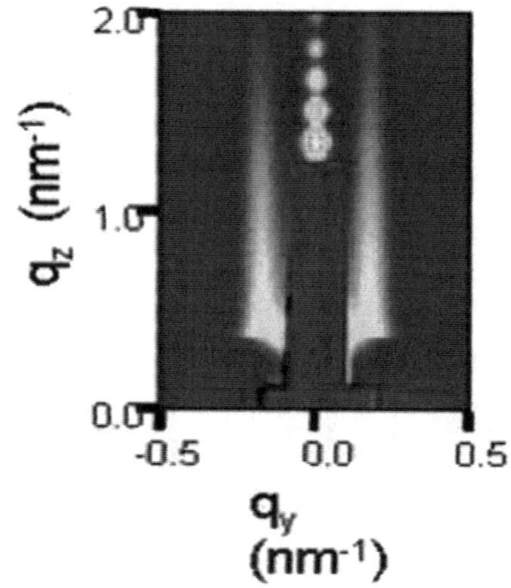

Figure 2. GISAXS pattern obtained for ϕ=45 degree with the angle of incidence α=0.35 degree. Two peaks extending towards q_z direction are from Ge nanodots. Rectangular shadow in the figure is the specular beam stop. The line of spots at q_y=0 extending towards larger q_z from the specular beam stop is a diffuse part of Kiessig fringe of the layer structure.

present GISAXS profiles, we may conclude that the GISAXS from Ge nanodots appearing between the specular beam stop and $|q_y|$=0.2 and the diffuse scattering from surface/ interface roughness, appearing at q_y=0 and above the specular beam stop, are well separated. Therefore, we may regard that the scattering intensity at the larger q_y corresponds to the scattering from Ge nanodots. As shown in the previous report [12], effect of correction term from DWBA should be small at the large q_z. Figure 3 gives an example of GISAXS pattern and corresponding two-dimensional intensity profiles obtained by Born approximations. For example, the angle of incidence, α, used in the present measurements lies between 0.35 and 0.48 degree. Comparing the simulation results with these conditions, the correction term should be less than 1% considering the reflectivity at the angle. Then the size and shape of the nanodots were successfully reconstructed using these fitting methods. The shape parameters obtained from these fitting was R_y =8.8 nm in the base diameter and 2.3 nm in height, with a flat dome shape. These results agreed with direct observation using cross-sectional transmission electron microscope. Therefore, it is concluded that for size and shape analysis where we may use the scattering intensity at the region rather far from the Yoneda region to avoid dynamical effect, analysis under Born approximation gives enough information. This condition depends strongly on the size of nanodots. If the height of nanodots is large, then the characteristic q_z to be used for the analysis becomes too close to the critical angle, and dynamical effect need to be taken into account.

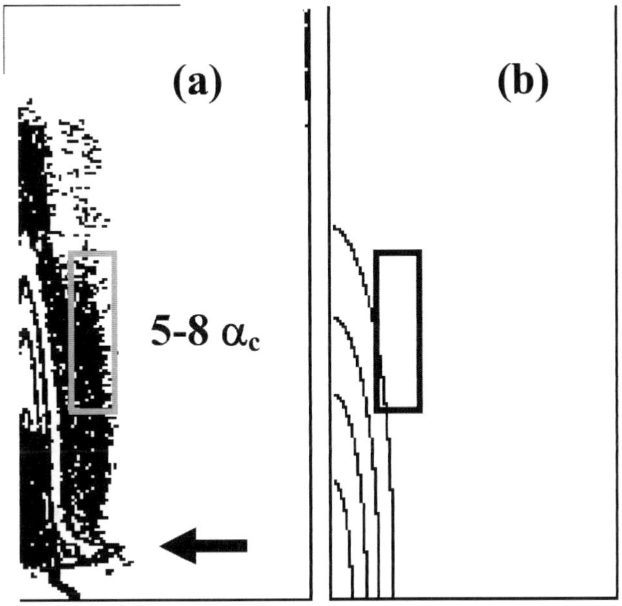

Figure 3. Measured two-dimensional GISAXS pattern contour (a) and a form factor of flat dome obtained by Born approximation (b). To fit the pattern for shape analysis, area of reference was selected where qz is large enough, and also q_y is larger than the interparticle interference peak to avoid the effect of structure factor. The fitting was made with a constraint that the gyration radius agrees with the experimental one. Naturally, dynamical effect is not negligible near the Yoneda line as shown by an arrow.

In the preceding analysis, we justified BA analysis from reflectivity curve shown in Fig.1 [9][10]. To make the analysis more quantitative, a DWBA simulation on the same sample was made. For the present sample, a simple layer model was used. For capped single layer of nanodots, GISAXS scattering amplitude is given in a form of [13]:

$$A(k_z^i, k_z^f, q_{//}) = T_i(k_z^i)T_f(k_z^f)A(k_z^f - k_z^i, q_{//}) + R_i T_f A(k_z^f + k_z^i, q_{//})$$

$$+T_i R_f A(-k_z^f - k_z^i, q_{//}) + R_i R_f A(-k_z^f + k_z^i, q_{//}) \quad (1)$$

with k_z^i, k_z^f, $q_{//}$ are the wave vector of incoming/outgoing perpendicular and in-plane wave, T and R are the transmission and reflection coefficients determined for the incoming/outgoing waves from recursion equation obtained for the reflectivity shown in Fig.1. From equation (1), the form factor of the scattering potential, namely, the shape of the nanodot, and the Fresnel coefficients as a function of angle are necessary for DWBA simulation of GISAXS patterns. As shown in Fig. 3, the form factor is obtained from BA analysis of the scattering intensity. GISAXS profile calculated for the present nanodot sample is calculated by :

$$I(q) = A(k_z^i, k_z^f, q_{//}) \cdot A(k_z^i, k_z^f, q_{//})^* \quad (2)$$

where A* is the complex conjugate of A.

In the present simulation, we assumed that the four terms appearing in equation (1) are coherent, i.e., the amplitude can be simply squared after summation to give the intensity. When the sample has some thick layer of scattering bodies, such as nanopores distributing over a relatively thick layer, the scattering occurred near the center of the layer and that reflected at the layer/substrate interface may not be necessarily coherent because the distance between the scattering body and the interface is large. For such cases some decoupling assumption is necessary [14]. In the present case, however, only one layer of nanodot are formed on a well-defined flat substrate. That is, the interface which gives the reflection in the DWBA analysis is adjoining to the dots. The four waves in equation 1 are expected to be therefore almost coherent.

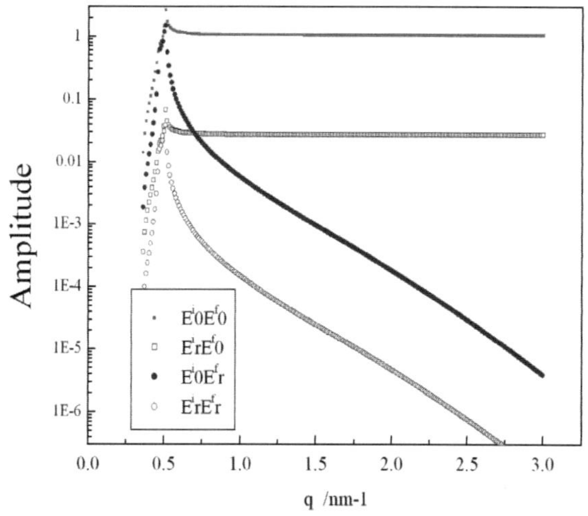

Figure 4. Magnitude of the Fresnel coefficient for the four components in the present condition. It is clearly seen that the magnitude of the third term is comparable to the first (Born) term near the Yoneda line.

Figure 4 gives the amplitude of the coefficients, corresponding to the factors to be multiplied to the form factor in equation 1 as a function of total moment transfer in q_z direction. The transmission part corresponding to the first term is the strongest in the all q_z region. In contrast, the transmission-reflection (T_iR_f) wave is quite strong near the Yoneda line, but decrease rapidly with q_z. Therefore, the effect of reflected beam, or DWBA correction term, is not negligible only in the region close to the Yoneda line. For example, the third term decreased by a factor of about 30 at the q_z just 0.2 nm^{-1} above the Yoneda line. Then, relative magnitude of the correction term to the total intensity is only 0.1%. Therefore, it is concluded that the effect of reflected beam (DWBA correction) is small in the present case and analysis based on Born approximation is sufficient when we can avoid the area in q space about 0.2 nm^{-1} within the Yoneda line. These considerations depend on the sample structures and the incident angle. Therefore, DWBA analysis might be still necessary to confirm the validity of the approximation.

GISAXS pattern calculated from these parameters described above is shown in Fig. 6, with two principal waves that contributes to the scattering intensity. For simplicity, the calculation is made for a single nanodot having a size that agrees with the average one obtained by experiment, and with the thickness and roughness of the cap layer given from the reflectivity simulation. It is seen that the GISAXS pattern shown in Fig.5(a) is well explained by the transmission(Born) term of Fig.5(b), except the region close to Yoneda line.

When the four waves in equation (1) are coherent, it is worthwhile looking at the relationship of the phases between the waves. Figure 6 gives the phase of the amplitude two waves, i.e., transmission wave(T_iT_f) and transmission-reflection (T_iR_f) waves shown in Fig. 5, which give principal contribution

Journal of Physics: Conference Series **184** (2009) 012005 doi:10.1088/1742-6596/184/1/012005

near the Yoneda line. The phase just on the Yoneda line does not show clear difference, because of the sign in the form factors in eq.(1). It also suggests that the intensity enhances just on Yoneda, and modulation occurs at the region very close to Yoneda lines. This might be important for a sample where a strong peak or line appears near the Yoneda line.

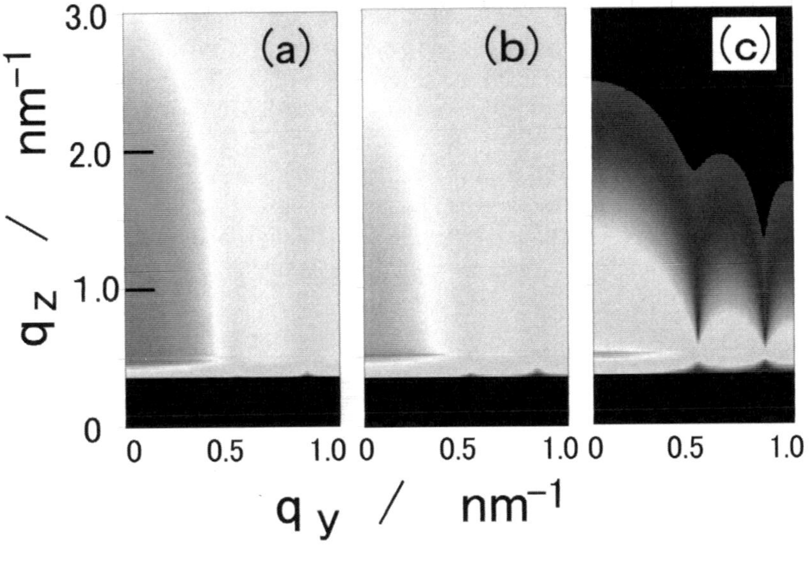

Figure 5. GISAXS intensity from single dome nanodot capped with a Si layer. (a) Total intensity with coherence assumed, (b) transmission wave (T_iT_f) intensity and (c) Reflection-Transmission (T_iR_f) waves. (b) and (c) gives principal contribution to GISAXS intensity near the Yoneda regions. It is also seen that GISAXS pattern shown in (a) is explained by the transmission term shown in (b). In contrast, R_iT_f term gives strong contribution only at the Yoneda line.

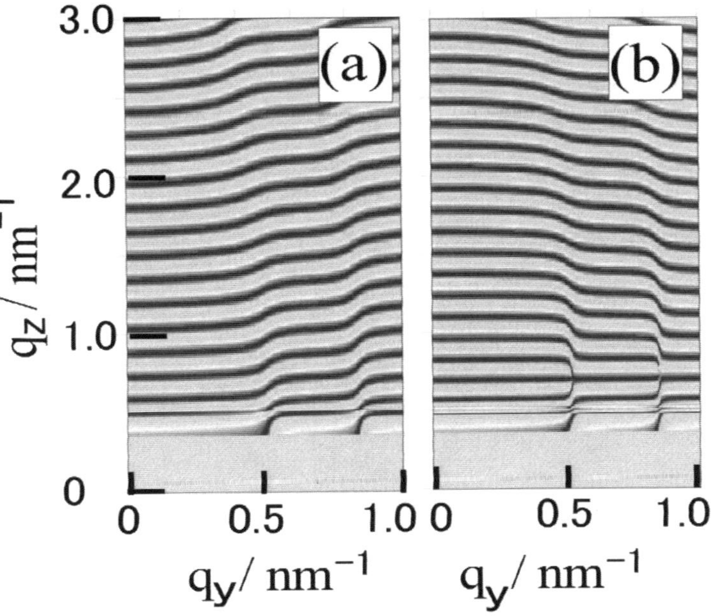

Figure 6. Argument of the complex amplitude of the waves for the first (a) and third (b) terms of equation (1) calculated for the GISAXS pattern shown in Fig. 5(b) and (c). It is seen that the pattern is reversed for the third term, as suggested from equation (1), implying that the phase pattern is reversed on reflection. The low q_z region below q_z=0.35nm^{-1} has no phase information because no scattering wave transmits here.

Journal of Physics: Conference Series **184** (2009) 012005 doi:10.1088/1742-6596/184/1/012005

4. Conclusions

Reflectivity/GISAXS measurements were performed at SPring8 and Photon factory for a Ge nanodot sample grown on Si (001) substrate and capped with Si layer by gas-source MBE. From a reflectivity analysis, parameters required for DWBA simulation of GISAXS were obtained. Detailed analysis of each wave component gives quantitative contribution of each wave in the GISAXS pattern formation. From Born approximation, size and shape of nanodots are evaluated. By using the form factor obtained by the BA and the layer structure information obtained by reflectivity, DWBA simulation was made.

It is concluded that Born approximation is sufficient to evaluate the microstructure of the present nanodot sample. However, it might not be the case for self-organized polymer films where incident angle is generally much smaller and structure is larger. In conclusion, a combined analysis of GISAXS /reflectivity is therefore a useful tool to reconstruct buried nanostructure near the surface, in particular to examine the validity of analysis.

References

[1] Sinha S.K., Sirota E.B., Garoff S.and Stanley H.B. 1988, *Phys. Rev. B* **38** 2297.

[2] Rauscher M. 1999 *J.Appl. Phys.* **86** 6763.

[3] Renaud G., Lazzari R., Revenant C., et al., 2003 *Science* **300** 1416.

[4] Paniago R., Metzger H., Rauscher M., Kovats, Z., Schultze J. Eisele I. and Ferrer S. 2000 *J.Appl. Cryst.*, **33** 433.

[5] Stangl J., Holy V.and Bauer G. 2004 *Rev. Mod. Phys.* **76** 725.

[6] Razzari L 2002 *J. Appl. Cryst.* **35** 406.

[7] http://www.insp.jussieu.fr/axe2/Oxydes/IsGISAXS/isgisaxs.htm

[8] Babonneau, D. Naudon, A., Thiaudiere D., and Lequien S. 1999 *J.Appl. Cryst.*,**32** 226.

[9] Okuda H., Ochiai S., Amemiya Y. and Ito K 2002 *Appl. Phys. Lett.* **81** 2358.

[10] Ogawa T., Niwa H. Okuda H.and Ochiai S., 2005 *Materials Science Forum*, **475-479** 1097.

[11] Tolan M., *'X-ray Scattering from Soft-Matter Thin Films'*, Springer Verlag Berlin1999.

[12] Okuda H., Ochiai S., Ohtaka M., Ichitubo T., Matsubara E., Usami N., Nakajima K., Sasaki S. and Sakata O., 2007*Trans. Mater. Res. Soc. Japan*, **32**275.

[13] Rauscher M., Salditt T. and Spohn H., 1995 *Phys. Rev B***52** 16855

[14] Lee, B., Yoon J. Oh W., Hwang Y.T. Heo K., Jin K.S., Kin J. Kim K-W. and Ree M.H. 2005 *Macromolecules* **38**, 3395.

Journal of Physics: Conference Series **184** (2009) 012006 doi:10.1088/1742-6596/184/1/012006

Dimensional and hydrodynamic properties of amylose tris(phenylcarbamate) in various solvents

Ken Terao

Department of Macromolecular Science, Osaka University, 1-1 Machikaneyama-cho, Toyonaka, Osaka 560-0043, Japan

E-mail: kterao@chem.sci.osaka-u.ac.jp

Abstract. Analyzes were made of the particle scattering function $P(k)$, z-average radii of gyration and intrinsic viscosities for amylose tris phenylcarbamate samples having different weight-average molecular weight M_w in 1,4-dioxane (DIOX), 2-ethoxyethanol (2EE), methyl acetate and 4-methyl-2-pentanone at 25 °C and in ethyl acetate at 33°C, on the basis of the wormlike cylinder model with or without excluded volume to determine the contour length per residue h and the Kuhn segment length. The h values in a ketone and esters were found to be significantly larger than those in DIOX and 2EE. Furthermore, the discrepancy between the experimental $P(k)$ for an ATPC sample (M_w = 18700) and the theoretical values at high k (absolute value of the scattering vector) region (k > 3 nm^{-1}) became more significant with decreasing h, suggesting that the $P(k)$ data at the k region reflect the helical structure of the polymer.

1. Introduction

In crystal structure, varieties of conformations having different helical pitch per residue h were found for amylosic chains; indeed, this h for amylose triesters ranges from 0.37 to 0.40 nm [1]. This indicates that the small perturbation (e.g. side chains and solvents) to the amylosic main chain may influence the local conformation and then h of amylosic chains in solution.

Much work has been done for dimensional and hydrodynamic properties of amylose tris phenylcarbamate (ATPC), whose chemical structure is shown in Figure 1, in solution [2–13]. This polymer has three pairs of polar NH and C=O groups. The above mentioned feature of amylosic chains suggests that h could depend on polar solvents having different chemical structure. However, we find only a few reports of h in solution. Thus we recently undertook studies on solution properties

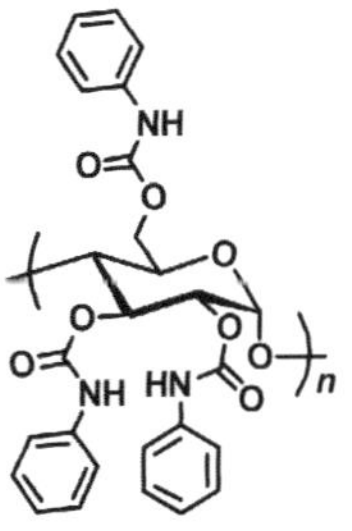

Figure 1. Chemical structure of amylose tris phenylcarbamate (ATPC).

of ATPC samples synthesized from enzymatically synthesized amylose [14]. To determine the weight-average molecular weight M_w, the z-average mean square radius of gyration $\langle S^2\rangle_z$, the particle scattering function $P(k)$ and the intrinsic viscosity $[\eta]$, where k denotes the magnitude of the scattering vector, light and synchrotron radiation small-angle X-ray scattering (SAXS) and viscosity measurements were made on the ATPC samples in 1,4-dioxane (DIOX), 2-ethoxyethanol (2EE), methyl acetate (MEA), ethyl acetate (EA) and 4-methyl-2-pentanone (MIBK). The obtained ratios of M_w to the number-average molecular weight or those of the z-average molecular weight to M_w were about 1.1 or less [15]. We made the measurements at 25 °C for DIOX, 2EE, MEA and MIBK solutions whereas 33 °C for the rest because it was determined to be the theta temperature. It is expected that carbonyl groups of ketones and esters may form hydrogen bonding with NH groups of ATPC, thus the solvent dependence in the local conformation of ATPC could be detectable as the difference of h.

This paper abstracts the major results from our recent analyses [15, 16] of the solution data for ATPC in terms of the Kratky-Porod (KP) wormlike chain model [17] and the possibility to estimate the local conformation of the polymer in solution from scattering data.

2. Results and discussion

2.1. Analysis in terms of the KP chain

Figure 2 shows the molecular weight dependence of $\langle S^2\rangle_z$ for ATPC in DIOX [15] and MIBK [16] at 25 °C. The discrepancy of $\langle S^2\rangle_z$ in the two solvents becomes more significant with decreasing M_w, indicating unmistakable difference of M_L (molar mass per unit contour length). The obtained $\langle S^2\rangle_z$, $P(k)$ and $[\eta]$ data in the five solvents were analyzed by the known theories [18, 19] for the KP chain model with excluded volume to determine M_L, λ^{-1}, the chain diameter d and the excluded-volume strength B; we note that one or two parameters of $[\eta]$ were assumed because all parameters cannot uniquely be determined. The theoretical solid curves for $\langle S^2\rangle_z$ (Figure 2) and $[\eta]$ (Figure 3) reproduce the corresponding experimental data almost quantitatively. Furthermore, the experimental

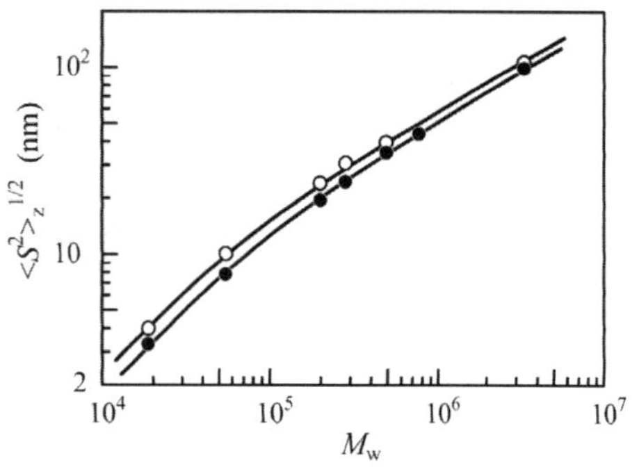

Figure 2. Molecular weight dependence of $\langle S^2\rangle_z^{1/2}$ for ATPC in DIOX (filled circles) [15] and MIBK (open circles) [16] at 25 °C. Solid curves, theoretical values for the perturbed wormlike chains with M_L = 1550 nm^{-1}, λ^{-1} = 24 nm, B = 1 nm in DIOX and M_L = 1190 nm^{-1}, λ^{-1} = 24 nm, B = 0.5 nm in MIBK.

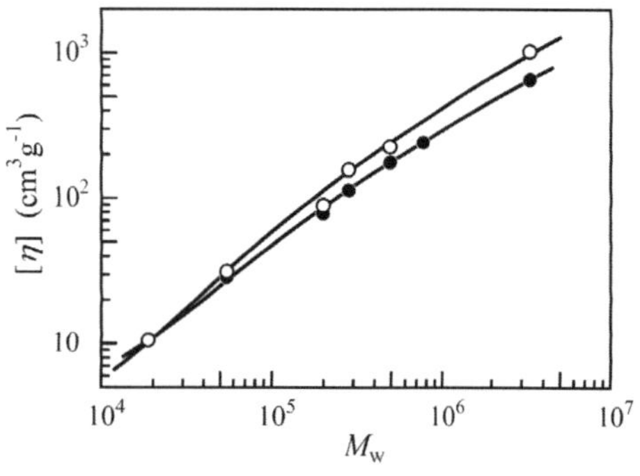

Figure 3. Molecular weight dependence of $[\eta]$ for ATPC in DIOX (filled circles) [15] and MIBK (open circles) [16] at 25 °C. Solid curves, theoretical values for the perturbed wormlike cylinders with M_L = 1540 nm^{-1}, λ^{-1} = 20 nm, d = 2.9 nm and B = 0.5 nm in DIOX and M_L = 1310 nm^{-1}, λ^{-1} = 23.5 nm, d = 1.9 nm and B = 0.5 nm in MIBK.

$P(k)$ for an ATPC sample (ATPC20K, $M_w = 1.87 \times 10^4$) is also fitted by the corresponding theoretical $P(k)$ in the range of $k < 3$ nm^{-1} (Figure 4); the discrepancy between the theoretical and experimental $P(k)$ in high k region reflects the local atomic structure of ATPC in solution. The molecular parameters obtained from the three different physical properties substantially agree within the possible experimental and determination errors other than d in DIOX; this inconsistency suggests the d value from SAXS can be affected by the electron density profile around the chain contour [18, 20]. The resultant λ^{-1} and h (contour length per repeat unit) calculated from $h = M_0/M_L$ (M_0: molar mass per repeat unit) are summarized in Table 1.

2.2. Solvent dependence of local chain length

The h values in DIOX and 2EE are significantly larger than those in a ketone and esters; i.e. h in MIBK is about 30% longer than that in 2EE, showing that the local conformation of ATPC significantly depends on the solvent. This behaviour is most likely due to the difference in the hydrogen bonding between intramolecular NH and C=O groups and intermolecular hydrogen bonding between the polar groups of ATPC and solvent molecules. We note that the intramolecular hydrogen bonding of ATPC in DIOX and 2EE was observed as splitting amide I band in their solution IR spectra [15]. We may thus speculate that the intramolecular hydrogen bonding is replaced in ketones and esters by the intermolecular hydrogen bonding between the NH groups of ATPC and the C=O groups of the solvent molecules, and then the main chain of ATPC extends. Slightly increase both h and λ^{-1} with molar volume of the solvent (MEA < EA < MIBK) seems to be consistent with this suggestion when we consider the hydrogen bonding solvent molecules instabilize the conformation

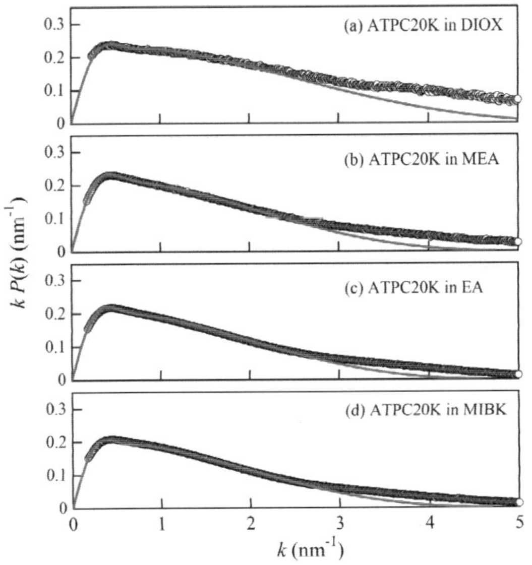

Figure 4. Holtzer plots for ATPC20K ($M_w = 1.87 \times 10^4$) in DIOX at 25 °C, MEA at 25 °C, EA at 33 °C, MIBK at 25 °C. Solid curves, theoretical values for the wormlike cylinder with $M_L = 1500$ nm^{-1}, $\lambda^{-1} = 21$ nm, $d = 1.3$ nm in DIOX, $M_L = 1380$ nm^{-1}, $\lambda^{-1} = 14$ nm, $d = 1.6$ nm in MEA, $M_L = 1310$ nm^{-1}, $\lambda^{-1} = 16$ nm, $d = 1.6$ nm in EA and $M_L = 1270$ nm^{-1}, $\lambda^{-1} = 23$ nm, $d = 1.6$ nm in MIBK. (The data in the range of $k < 2.5$ nm^{-1} have been published in refs [15, 16]).

Table 1. Comparison of h and λ^{-1} for ATPC in Various Solvents.

Solvent	T (°C)	h (nm)	λ^{-1} (nm)
2EE [a]	25	0.32 ± 0.01	16 ± 2
DIOX [a]	25	0.34 ± 0.01	22 ± 2
MEA [b]	25	0.37 ± 0.02	15 ± 2
EA [b]	33	0.39 ± 0.02	17 ± 2
MIBK [b]	25	0.42 ± 0.02	24 ± 2

[a] Reference [15]. [b] Reference [16].

having short helical pitch. However, no spectroscopic evidence has been obtained for either intra- or intermolecular hydrogen bonding in ketones and esters because of inapplicability of either FT-IR or circular dichroism measurements. Investigation of appropriate amylose derivative-solvent systems could be useful to clarify solvent dependent local conformational behaviour of amylosic chains.

2.3. Scattering data at high k range

The discrepancy between theoretical and experimental scattering function in Figure 4 occurs at the region $k > 3\ nm^{-1}$, indicating that the continuous cylinder model should not be applicable to the data there. Interestingly, the difference increases with decreasing h. This is reasonable when we consider the higher scattering intensity than the theoretical value reflects the local helical structure which becomes more appreciable with shortening h. Thus, the analyses of $P(k)$ at high k range are probably useful to determine the local conformational features of amylose derivatives in solution. However, the scattering function at the k range is generally affected by the local conformational fluctuation and the electron density profiles of solvent molecules around the polymers chains [20]; the former probably weakens the characteristic peaks. Therefore, no further analysis is made at the present time due to the difficulty of the estimation of each effect on $P(k)$. We may conjecture that the comparison between the atomic structure generated by molecular simulation and the accurate and wide-range experimental scattering functions is helpful to obtain new insights for the local conformation of the polymers in solution.

Acknowledgements

The author thanks Prof. T. Norisuye, Prof. T. Sato (Osaka University) and Prof. S. Kitamura (Osaka Prefecture University) for fruitful discussions. The synchrotron radiation experiments were performed at the BL40B2 in SPring-8 with the approval of the Japan Synchrotron Radiation Research Institute (JASRI) (Proposal #2007A1034, #2007B1296 and #2008A1313).

References

[1] Zugenmaier P and Steinmeier H 1986 *Polymer* **27** 1601
[2] Burchard W 1964 *Z. Physik. Chem.* **42** 293
[3] Cowie J M G 1965 *Biopolymers* **3** 69
[4] Burchard W 1965 *Makromol. Chem.* **88** 11
[5] Banks W, Greenwood C T and Sloss J 1970 *Makromol. Chem.* **140** 109
[6] Banks W, Greenwood C T and Sloss J 1970 *Makromol. Chem.* **140** 119
[7] Banks W and Greenwood C T 1971 *Makromol. Chem.* **144** 135
[8] Banks W, Greenwood C T and Sloss J 1971 *Eur. Polym. J.* **7** 879
[9] Burchard W 1971 *Br. Polym. J.* **3** 214
[10] Sutter W and Burchard W 1978 *Makromol. Chem.* **179** 1961
[11] Gupta A K, Marchal E, Burchard W and Pfannemüller B 1979 *Macromolecules* **12** 281
[12] Pfannemüller B, Schmidt M, Ziegast G and Matsuo K 1984 *Macromolecules* **17** 710
[13] Muroga Y, Hayashi K, Fukunaga M, Kato T, Shimizu S and Kurita K 2006 *Biophys. Chem.* **121** 96
[14] Kitamura S, Yunokawa H, Mitsuie S and Kuge T 1982 *Polym. J.* **14** 93
[15] Terao K, Fujii T, Tsuda M, Kitamura S and Norisuye T 2009 *Polym. J.* **41** 201
[16] Fujii T, Terao K, Tsuda M, Kitamura S and Norisuye T 2009 *Biopolymers* in press DOI: 10.1002/bip.21219
[17] Kratky O. and Porod G 1949 *Recl. Trav. Chim. Pays-Bas* **68** 1106
[18] Yamakawa H 1997 *Helical Wormlike Chains in Polymer Solutions* (Berlin: Springer)
[19] Nakamura Y and Norisuye T 2004 *J. Polym. Sci. B: Polym. Phys.* **42** 1398
[20] Terao K, Mizuno K, Murashima M, Kita Y, Hongo C, Okuyama K, Norisuye T and Bächinger H P 2008 *Macromolecules* **41** 7203

Journal of Physics: Conference Series **184** (2009) 012007 doi:10.1088/1742-6596/184/1/012007

Influence of α-methyl group on molecular aggregation structure and surface physicochemical properties of fluoroalkyl side chain polymers

K. Honda[1], H. Yamaguchi[1], O. Sakata[2], S. Sasaki[2,3], M. Takata[2,3], M. Morita[4], A. Takahara[1,5]

[1]Graduate School of Engineering, Kyushu University, Motooka, Nishi-ku, Fukuoka, 819-0395, Japan, [2]Japan Synchrotron Radiation Research Institute (JASRI) / SPring-8, Hyogo 679-5198, Japan, [3]The RIKEN Harima Institute / SPring-8, Hyogo 679-5198, Japan, [4]Chemical Division, Daikin Industries, Ltd.,1-1 Nishi Hitotsuya, Settsu-shi, Osaka 566-8585, Japan, [5]Institute for Materials Chemistry and Engineering, Kyushu University, Motooka, Nishi-ku, Fukuoka, 819-0395, Japan

E-mail: takahara@cstf.kyushu-u.ac.jp

Abstract. Influence of α-methyl group on molecular aggregation states and surface physicochemical properties of poly(fluoroalkyl acrylate)s [PFA-C_y, where y is fluoromethylene number in R_f group] and poly(fluoroalkykl methacrylate)s [PFMA-C_y] thin films were systematically investigated. Spin-coated PFA-C_y and PFMA-C_y thin films were characterized by dynamic contact angle measurements and grazing-incidence wide-angle X-ray diffraction (GIWAXD) measurements. GIWAXD data revealed that fluoroalkyl side chains of PFA-C_y and PFMA-C_y with $y \geq 8$ formed regular structures in the surface region as well as bulk one. However, the degree of orientation and ordering of the R_f groups of PFMA-C_8 thin films was low. Also, the receding contact angle (θ_r) of PFMA-C_8 thin films was lower than that of PFA-C_8 ones. By annealing treatment, the θ_r of PFMA-C_8 was increased. These results suggest that the R_f groups of PFMA-C_8 were disordered due to presence of the α-methyl group. The R_f groups became ordered to pack closely each other by annealing treatment, so that the water repellency was increased.

1. Introduction

Polymers with long fluoroalkyl (R_f) groups have high water and oil repellencies and they are widely used for surface modifiers. It is well known that the surface molecular aggregation states and physicochemical properties of these polymer depend on side chain R_f groups [1-4]. In this paper, The surface molecular aggregation structure and physicochemical properties of poly(fluoroalkyl acrylate) [PFA-C_y, where y is the fluoromethylene number of R_f groups] and poly(fluoroalkyl methacrylate) [PFMA-C_y] thin films were investigated, and influence of the α-methyl group on surface molecular aggregation state and water repellency was discussed.

2. Experimental

2.1. Material. The chemical structure of PFA-C_y and PFMA-C_y is shown in Figure 1. Polymers were prepared by free radical polymerization. PFA-C_y and PFMA-C_y thin films were prepared by the spin-coating method (2000 rpm, 30 s). The film thickness was estimated as approximately 100 nm by atomic force microscopy (AFM).

2.2. Wide-angle X-ray diffraction (WAXD) measurements. The WAXD measurements were carried out for powder samples on a RINT 2500V (Rigaku Co., Ltd.) with a Cu-K_α X-ray source (40 kV, 200 mA). The wavelength, λ, of the incident X-ray was 0.1542 nm. The scattering vector, q, was defined as $(4\pi/\lambda)\sin\theta$.

2.3. Grazing-incidence wide-angle X-ray diffraction (GIWAXD) measurements. GIWAXD measurements were carried out at 300 K with a four-circle diffractometer installed at a BL-13XU beamline of SPring-8 (JASRI, Japan). The λ of incident X-rays used in this study was 0.1284 nm or 0.1025 nm. The data collection time was 1.5 sec per step, and the angular interval between steps was 0.05°. Figure 2 shows the schematic geometry of the in-plane and out-of-plane GIWAXD measurement. Bragg diffraction from the surface region of a sample was obtained at α_i = 0.10° or 0.08° below the critical angle of total external reflection at the surface of the sample [5]. In the in-plane geometry, scattering vector (q_{xy}) is parallel to the sample surface, and the detected profiles show crystallographic regularity of the lattice plane being perpendicular to the film surface. On the other hand, information about crystallographic regularity of the lattice plane being parallel to the surface is obtained from out-of-plane geometry.

2.4. Dynamic contact angle measurement. Wetting properties were evaluated by dynamic contact angle measurements. The dynamic contact angle was measured using an inclinable plane [6]. A sample on a stage was tilted from an angle where the sample surface is level with the ground until a 50 μL water droplet began to slide down along the sample surface. Subsequently, an advancing contact angle (θ_a), a receding contact angle (θ_r), and a sliding angle (θ_s) were determined.

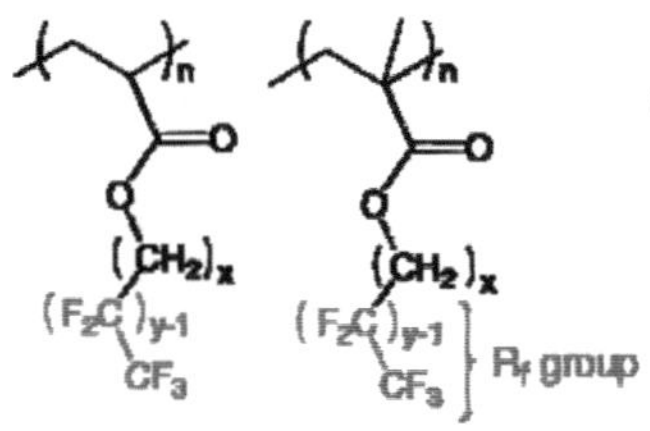

Fig.1 Chemical structures of PFA-C_y and PFMA-C_y (x = 1 for 1 and 2, x = 2 for y ≥ 4).

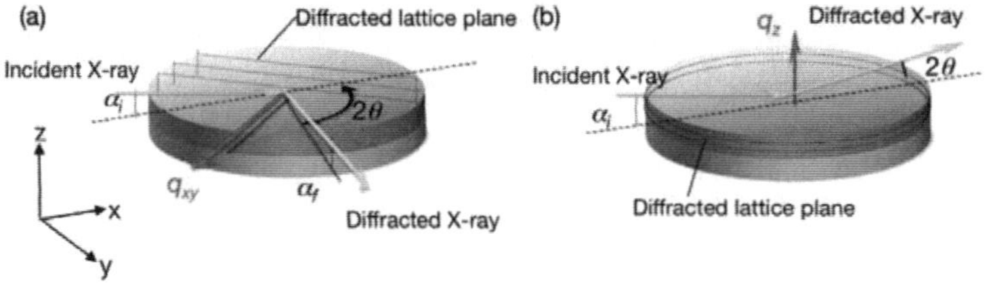

Fig.2 Schematic geometry of (a) in-plane and (b) out-of-plane GIWAXD measurement.

3. Result and Discussion

3.1. WAXD measurements. Fig.3 shows WAXD profiles of (a) PFA-C_y and (b) PFMA-C_y. PFA-C_y and PFMA-C_y with y ≤ 6 had no sharp peak. For those with y ≥ 8, Diffraction peaks were measured at q = 1.0 – 7.0 nm^{-1} and ca. 12.3 nm^{-1}. The peaks at q = 1.0 – 7.0 nm^{-1} were explained as reflections from the lamellar structure in which R_f groups were ordered like a multilayered structure (the *d*-spacing [3.4 – 3.9 nm] was in good agreement with length of the two R_f groups) [7]. The other peaks at q = ca. 12.3 nm^{-1} were assignable to reflections from a regular packing structure of the R_f groups (d = ca. 0.50 nm, which is close to the intermolecular distance of the PTFE crystal [d = 0.49 nm], which shows *pseudo*-hexagonal packing of fluoroalkyl chains) [8]. Thus, PFA-C_y and PFMA-C_y with $y \geq 8$ are in the crystalline state. Then, the molecular aggregation structure in the near-surface region of PFA-C_8 and PFMA-C_8 thin films was measured by GIWAXD measurements.

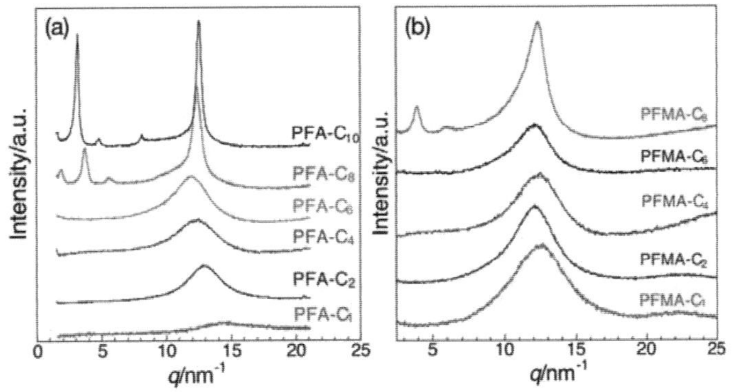

Fig.3 Powder WAXD profiles of (a) PFA-C_y and (b) PFMA-C_y. The λ of incident X-ray was 0.1542 nm. The data collection time was 3 s per step, and the angular interval between steps was 0.05°.

3.2. GIWAXD measurements. Fig.4 show the (a) in-plane and (b) out-of-plane GIWAXD profiles of PFA-C_8 and PFMA-C_8 thin film surfaces. In in-plane and out-of-plane GIWAXD profiles of PFA-C_8 and PFMA-C_8 samples, diffractions from packing structures of R_f groups and lamellar structures of these polymers were clearly observed, respectively. These results indicated that the R_f groups were oriented perpendicular to the film surface and stacked each other parallel to it. The lamellar domains were stacked perpendicular to the film surface. Then, the GIWAXD profiles were compared with annealed one (PFA-C_8: at 333 K and 348 K (T_m) for 6 h, PFMA-C_8: 333 K, 348 K, and 358 K (T_m) for 6 h). For PFA-C_8, the peak attributed to the lamellar structure was also observed in an in-plane GIWAXD profile before annealing. However, this peak became weaker in intensity for a PFA-C_8 annealed at 333 K and disappeared for one annealed at 348 K. Another peak attributed to the *pseudo*-hexagonal packing of the R_f groups became sharper with an increase in annealing temperature. By annealing treatment, some R_f groups changed their orientation from the random orientation to the perpendicular one against the film surface. In the case of out-of-plane GIWAXD, diffractions from the lamellar structure became sharper in peak profile with an increase in annealing temperature. In the case of PFMA-C_8, diffractions from the packing structures of the R_f groups and the lamellar structures also became sharper in peak profile and stronger in intensity by annealing treatment. However, diffractions attributed to *pseudo*-hexagonal packing of the R_f groups were broader in peak profile and weaker in intensity than those of PFA-C_8. This indicates that the small size or less ordered orientation of the packed R_f groups was formed for PFMA-C_8 compared with PFA-C_8. This structural disordering of R_f groups of PFMA-C_8 probably relates to the presence of the α-methyl group that restricted the orientation of R_f side chains. This behavior is similar to that observed for poly(alkyl methacrylate)s [9].

3.3. Dynamic contact angle measurement. To evaluate the effect of these molecular aggregation states on wetting properties, the dynamic contact angle measurement was carried out. Table 1 summarized the results of dynamic contact angle against water for PFA-C_8 and PFMA-C_8 thin film before and after annealing. For PFA-C_8, the value of θ_r was high and there were no differences in the dynamic contact angle before and after annealing. On the other hand, the θ_r of PFMA-C_8 before annealing was lower than that of PFA-C_8 and increased after annealing. In the case of PFA-C_8, the orientation and ordering of R_f groups before annealing was sufficiently high for achieving high water repellency. For the PFMA-C_8, the degree of orientation and ordering of R_f groups before annealing was extremely low. As a result, θ_r of PFMA-C_8 before annealing was lower than that of PFA-C_8. After annealing, the R_f groups became ordered closely and the water repellency was increased.

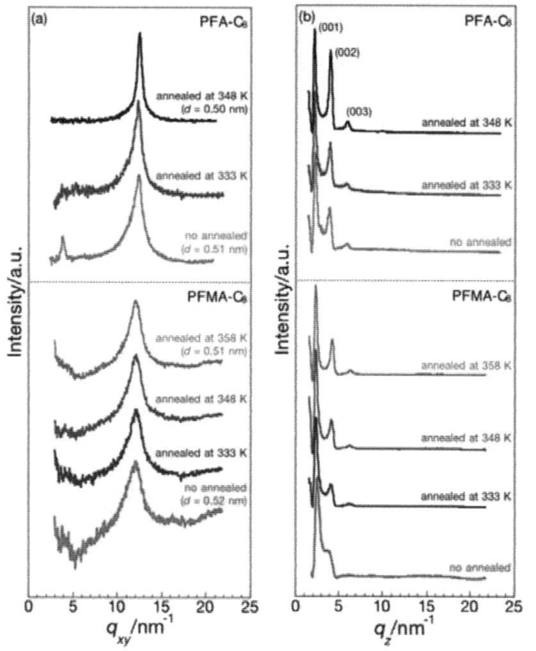

Fig.4 (a) in-plane and (b) out-of-plane GIWAXD profiles measured at surface region of PFA-C_8 (upper) and PFMA-C_8 (bottom) thin films. The λ of incident X-ray was 0.100 nm.

Table 1. Dynamic contact angle against water (50 μL) for PFA-C_8 and PFMA-C_8 thin

	Dynamic contact angle		
	θ_a/deg.	θ_r/deg.	α/deg.
PFA-C_8	124 ± 1.1	108 ± 0.6	12 ± 1.2
PFA-C_8 (annealed)	124 ± 0.6	112 ± 0.5	11 ± 0.6
PFMA-C_8	123 ± 2.1	90 ± 2.1	22 ± 3.1
PFMA-C_8 (annealed)	122 ± 1.3	112 ± 1.1	7 ± 1.2

4. Conclusions

Influence of the α-methyl group on molecular aggregation structure and surface physicochemical properties of PFA-C_y and PFMA-C_y thin films were systematically investigated as a function of the fluoromethylene number of side-chain R_f groups, y. WAXD measurement revealed that the PFA-C_y and PFMA-C_y with $y \geq 8$ formed crystal structures. GIWAXD revealed that fluoroalkyl side chains of PFA-C_y and PFMA-C_y with $y \geq 8$ formed regular structures in the near-surface region as well as in bulk one. However, the R_f groups of PFMA-C_8 thin films formed less ordered small size crystallite. The θ_r of a PFMA-C_8 thin film was lower than that of a PFA-C_8 thin film. By annealing treatment, the θ_r of the PFMA-C_8 thin film became increased. These results indicated that crystallographic disordering of the R_f groups of the PFMA-C_8 might relate to the aggregation state of α-methyl group, and the R_f groups became ordered and the water repellency was improved by annealing treatment.

Acknowledgements
The synchrotron X-ray diffraction experiments were performed at the BL13XU in the SPring-8 with the approval of the Japan Synchrotron Radiation Research Institute (JASRI) (Proposals No.2007A1429 & No. 2007B1530).

References

[1] Honda K, Morita M, Otsuka H and Takahara A 2005 *Macromolecules* **38** 5699.
[2] Honda K, Morita M and Takahara A 2007 *Kobushi Ronbushu* **64** 181.
[3] Honda K, Yamaguchi H, Kobayashi M, Morota M, Takahara A 2008 *J. Phys. Conf. Ser.* **100** 012035.
[4] Honda K, Morita M, Takahara A 2009 *Macromolecules* submitted.
[5] Russell T P 1990 *Mater. Sci. Rep.* **5** 171.
[6] Extrand C W and Kumagai Y 1997 *J.Colloid Interface Sci.* **191** 378.
[7] Volkov V V, Plate N A, Takahara A, Kajiyama T, Amaya N and Murata Y 1992 *Polymer* **33** 1316.
[8] Bunn C W and Howells E R 1954 *Nature* **174** 549.
[9] Rogers S S, Mandelkern L 1957 *J. Phys. Chem.* **61** 985.

Journal of Physics: Conference Series **184** (2009) 012008 doi:10.1088/1742-6596/184/1/012008

Supramolecular Structures of Benzyl Amine Derivate/DNA Complexes Explored with Synchrotron Small Angle X-ray Scattering at SPring-8

Mina Sakuragi[1], Shouta Kusuki[1], Emi Hamada[1], Hiroyasu Masunaga[2], Hiroki Ogawa[2], Isamu Akiba[1], and Kazuo Sakurai[1]*

[1]Department of Chemistry and Biochemistry, Faculty of Engineering, The University of Kitakyushu, 1-1, Hibikino, Wakamatsu-ku, Kitakyushu, Fukuoka 808-0135, Japan,
[2] JASRI/SPring-8

sakurai@env.kitakyu-u.ac.jp

Abstract. Relationship between the transfection efficiency and the supramolecular structure of the lipoplexes consisting of DNA, a benzyl amine derivative (BA), and two neutral co-lipids (DOPE and DLPC) were examined. At a composition of BA:DOPE:DLPC = 1:2:1; where the highest efficiency was attained, SAXS showed that the micelles before adding DNA formed a hexagonally-packed cylinder and addition of DNA improved the packing order. This phenomenon may be interpreted by intercalation of DNA between the preformed micellar cylinders. At a composition of BA: DLPC=1:1, where the lowest efficiency was observed, the micelle adopted a spherical form and addition of DNA induce a structural transition from the sphere to a hexagonally packed cylinder, and then a lamellar form. Fitting with a multilayer model suggested that DNA was included into the hydrophobic layer of the complex. Comparison of these two compositions suggests that difference in the DNA location in the lipoplex is related to the transfection efficiency.

1. Introduction

The supermolecular complexes made from DNA and synthetic cationic lipids, called “lipoplex”, a coined term from the combination of “liposome” and “complex”, attracted a large amount of scientific and technological interest ever since the pioneering work done by Felgner *et al* [1]., that cationic lipids can be used to transfer DNA into mammalian cells. One of the major reasons to invoke such attention is that they can replace viral vectors and thus possibly resolve safety issues inherent for them. Viral vectors are prepared by genetically modifying a highly infectious natural virus and the safety issues have been concerned in terms of immunity as well as possibility of genome-integration. Although suppliers do not disclose the chemical structure and ingredients, several types of cationic lipids are widely used as a transfection regent for *in vitro* studies in these days. However, they have still difficulties in extending *in vivo* use owing to low transfection and unknown mechanism of their expression.

Safinya *et al* [2]. are the first to investigate relation between the supramolecular structures of lipoplexes and their transfection efficiency with synchrotron small angle X-ray scattering (SAXS) and to show the efficiency is strongly related to lipoplex’s structures. According to them, there are two predominant phases, (a): multilamellar phase where DNA monolayers are intercalated between lipid bilayers (L_α^C) and (b): the inverted hexagonal phase with DNA encapsulated within cationic lipid monolayers tubes and arranged on a two-dimensional hexagonal lattice (H_{II}^C). It is believed that the later structure is more suitable for the transfection, because of the fact that inverted hexagonal structures are often observed in the region where biological membranes fuse [3,4].

Independently from these previous studies, our group reported that aromatic amine and amidine compounds can be used as a transfection regent with a better efficiency and less toxic than commercial products [5]. Figure 1 presents the gene transfection efficiency for our benzyl amine (BA) system as a function of the composition, where the efficiency was evaluated with an established method called "luciferase assay" [6] which spectroscopically measures the amount of the light-producing protein expressed by foreign plasmid DNA (pDNA). The efficiency strongly depends on the composition: the highest one is achieved at BA: DOPE: DLPC=1: 2: 1 (point B) and lowest ones are near the A point (BA: DOPE: DLPC = 1: 0: 1), where DOPE and DLPC are L-a-phosphatidyl ethanolamine dioleoyl and 1,2-dilauroyl-sn-glycero-3-phosphocholine, respectively. The coexisting DOPE and DLPC are considered to make BA to be compatible with water and to be less toxic for cells. However, there has been no information how these two colipids affect the micellar structure. In this paper we compares the supramolecular structures between A and B with SAXS.

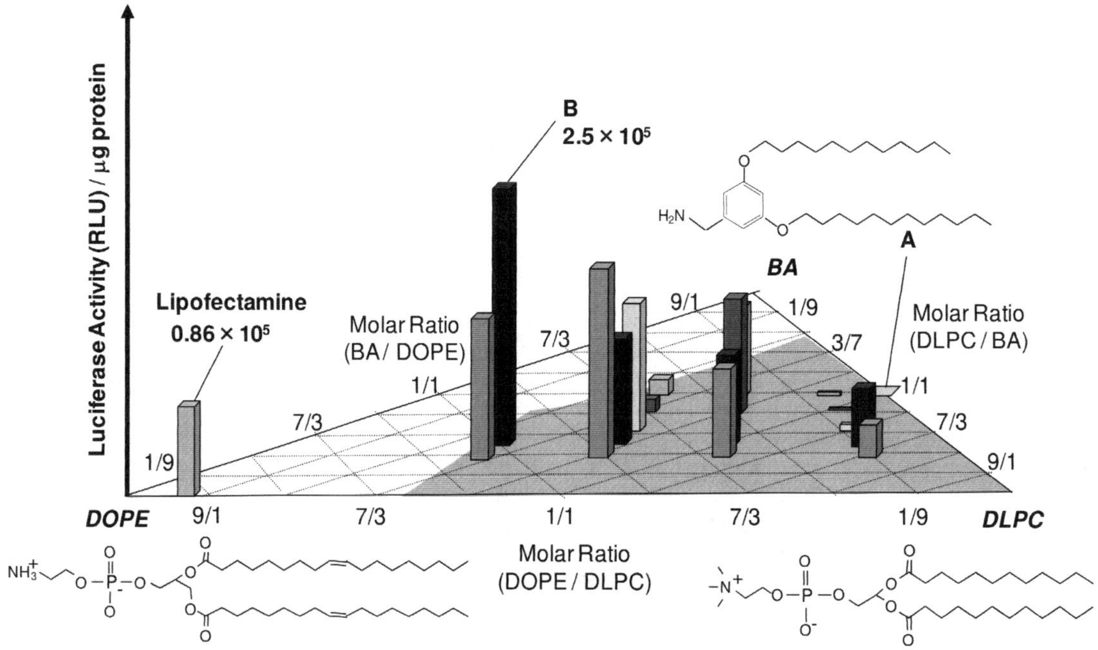

Figure 1. Composition dependence of the luciferase activity for the BA/DOPE/DLPC system. N/P ratios were fixed at 3.3 for all samples.

2. Experiments and Fitting Equations

Material: DOPE and DLPC were purchased from Wako chemical and Cosmopolitan biology, respectively, and BA was synthesized with a reported method[5]. BA, DOPE and DLPC were dissolved in chloroform at a certain composition and freeze-dried in vacuum. The micelle solutions were prepared by adding a 50 mM NaCl solution to the freeze-dried lipid mixture and subsequently dispersed by ultra-sonication. Plasmid DNA coding pGL3 (bp = 5256) was amplified with *E. coli* competent cells and purified by using a QIAGEN HiSpeed Plasmid Maxi kit. The micellar solutions and pGL3 were mixed and stirred with a vortex for 40 s, about 20 minutes before the measurements. The micelle/DNA composition was expressed with N/P that is defined by the molar ratio of the nitrogen of BA to the phosphate of DNA.

SAXS Set-Up: SAXS measurements were carried out at 40B2 SPring-8 with a 1.5 m camera using a Rigaku imaging plate (30×30 cm, 3000×3000 pixels) as a detector. The wavelength of the beam was 1.0 or 0.72 Å, the exposure time was 5 min, and the relative X-ray intensity and the sample transmittance were determined with two ion-chambers located in front of and behind the sample. A bespoke chamber for SAXS measurements was constructed that enables a sample to be introduced in vacuum and scattering measured from it, as well as a vacuum-proof cell with a 0.10 mm quartz glass plate [7]. Combining these two instruments, we improved the S/N ratio by almost ten times in the entire *q* range and eliminated parasitic scattering, compared with conventional set-up, where *q* is the magnitude of the scattering vector. One of the advantages of this vacuum set-up is to enables to measure SAXS from the solutions with the same dilute conditions as those for bio assays.

Fitting Models and Electron Density Calculation:

The form factor for multilayer spherical (P_S) or cylindrical micelles (P_C) is given by Eq 1 [8].

$$P_S(q) or P_C(q) \propto \frac{1}{q^a}\sum_{i=1}^{k}\left[\frac{V_i(\rho_i-\rho_{i+1})F_1(qr_i)}{qr_i}\right]^2 \quad \begin{matrix} a=1 \text{ for cylinders} \\ a=0 \text{ for spheres} \end{matrix} \qquad (1)$$

Here, F_1 is the first order Bessel function for cylinders or the first order spherical Bessel function for spheres, respectively. ρ_i, V_i, and r_i are the electron density of the i-th layer, the volume of the i-th layer, and the radius of the i-th layer, respectively. In Eq 1, ρ_{k+1} is the electron density of the solvent. When applied to the data, we assumed the Gaussian distribution for the radius. As the structural factor for the hexagonal packing [$S_h(q)$], we used the theoretical formula obtained by Hashimoto et al [9] that contains the paracrystalline distortion factor g. The electron density of the solvent, core, and shell were calculated from the following equation;

$$\rho = d\sum_i \rho_{w,i} w_i \qquad (2)$$

Here, d is the density of the mixed solvent, $\rho_{w,i}$ where w_i is the electron density per weight and the weight fraction for the i component, respectively.

3. Results

3-1: Structural Transition at Point A

Figure 2a presents N/P dependence of SAXS at the composition of A (the lowest transfection efficiency). The SAXS profile before adding DNA can be characterized as a multilayered spherical micelle and thus the data was fit with the corresponding model (a = 0 and k = 2 in Eq 1). Here, we adopted three fitting parameters: r_1, r_2, and ρ_2 (the radii of the core and shell, and the electron density of the shell, respectively) under the condition that the core and solvent electron densities were fixed at $\rho_1 = 210$ e/nm^3 and $\rho_3 = 334$ e/nm^3, respectively. These densities were calculated from Eq 2 and can be treated as a constant with respect of other three parameters [10]. The obtained parameters are summarized in Table 1 and the best fit curve is shown as a solid line in Figure 2a, indicating that the model can well reproduce the experiment qualitatively. Based on these parameters, Figure 3a depicts the inner struc erms of the

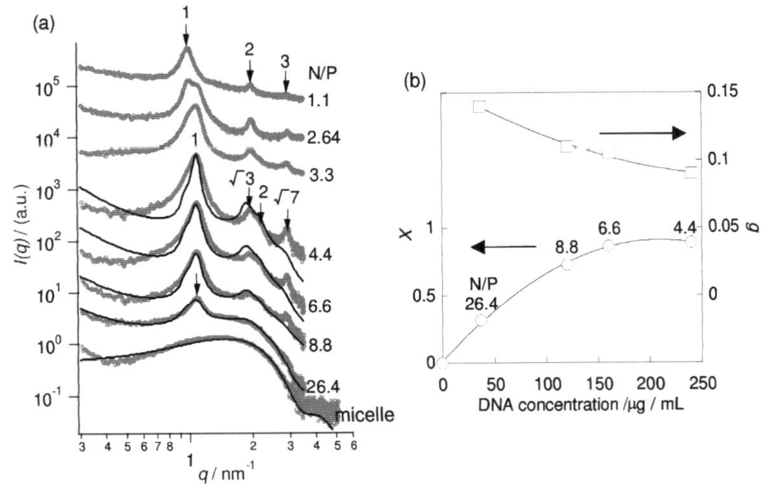

Figure 2. (a) N/P dependence of SAXS profile at the composition A (BA:DLPC=1:1). The micelle concentrations were 3.2 mM at N/P=3.3 - 26.4, 2.56 mM at N/P=2.64, and 1.07 mM at N/P=1.1. (b) N/P dependence of g (□) and 1-x (○).

chemical structure of the lipids. The electron density of the shell (ρ_2) is larger than the values estimated from the chemical structure, suggesting that the salt molecules are condensed on the head-groups.

When DNA was added to the micelle, a diffraction peak appeared at q = 1.1 nm^{-1} at N/P = 26.4, and, in the range of N/P>4.4, further addition caused to appear and increase other peaks at q = 1.09, 1.92, 2.18, and 2.89 nm^{-1}, as well as decrease of the convex curve which represents micellar scattering. These peak-positions satisfied the relation expected for hexagonally packed cylinders and the N/P dependence indicates that all of the spherical micelles disappeared and only cylinders existed at N/P = 4.4. This fact implies that the conversion from the DNA-un-bound spheres to the DNA-bound cylinders was seemed to complete at N/P = 4.4. If this hexagonally packed cylinder takes H_{II}^{C} (see the introduction), most of the lipids have to be bound to DNA unless the complex size is comparable with that of the cylinders [6]. We measured dynamic light scattering from the complex solution and analyzed the data with cumulant method [11]. The data showed that the size of the complex (c.a., 117 nm in diameter) was much larger than the cylinders, it can be concluded that there are almost 8 lipid molecules per one DNA phosphate anion from the relation of BA: DOPE: DLPC = 1: 0: 1 and N/P = 4.4. This lipid/DNA molar ratio indicates that H_{II}^{C} seems unlikely for the present system, because H_{II}^{C} requires one-to-one interaction between the phosphate and the cationic head. We fit the profile at N/P = 4.4 with $P_C(q)\ S_h(q)$, where $P_C(q)$ is the cylinder's form factor and $S_h(q)$ is the hexagonal structural factor, assuming only hexagonally packed cylinders existing. The diameter of the cylinder is larger than that of the micelle and the cylinder contains a central core with the almost same electron density

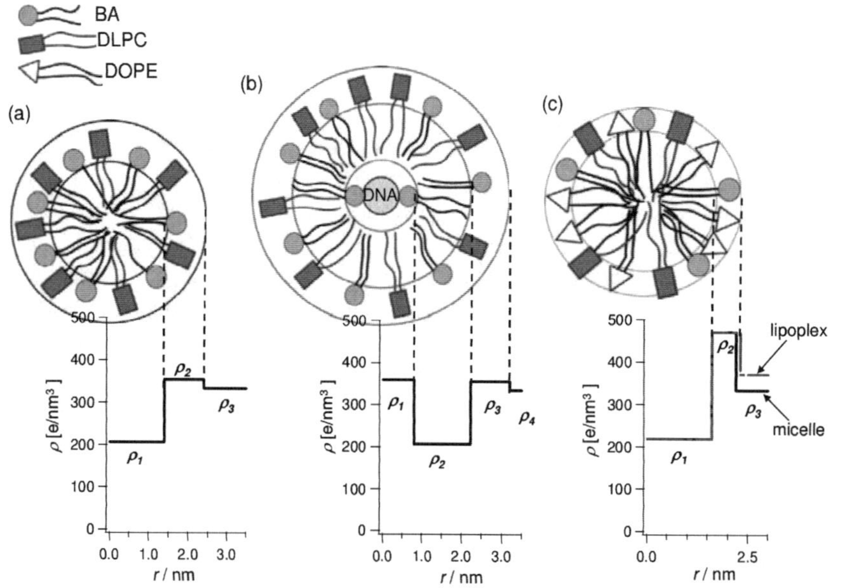

Figure 3. (a) The core-shell sphere model used to fit the micellar data and its electron density profile, (b) The three layered cylinder model used to fit the profiles at N/P = 4.4 in Figure 2a, and (c) The core-shell cylinder model used to fit all the profiles in Figure 4. We used different values for the solvent density: 334 e/nm^3 for the micelle and 370 e/nm^3 for the lipoplex.

Table 1. Structural Parameters for A or B

	structure	r_1 [nm]	r_2 [nm]	r_3 [nm]	ρ_1 [e/nm^3]	ρ_2 [e/nm^3]	ρ_3 [e/nm^3]
A-micelles	sphere	1.4	2.4	-	206	355	-
A-lipoplexes	cylinder	0.8	2.2	3.2	360	206	356
B-micelles	cylinder	2.2	1.6	-	220	470.5	-
B-lipoplexes	cylinder	2.3	1.6	-	220	470.5	-

as that of the shell. This result suggests that once DNA formed an ion-pair with BA, the DNA must be included inside of the cylinder and the cylinder surface is covered with the rest of the free lipids, as illustrated in Figure 3b. This model can rationalize the lipid/DNA molar ratio more reasonably than H_{II}^{C} model. Since the solutions at 26.4<N/P < 4.4 may be a mixture of the multilayered sphere given by Figure 3a and the hexagonally packed cylinder given by Figure 3b, the scattering profiles was fit by assuming the following relation:

$$I(q) = x_1 P_S(q) + x_2 P_C(q) S_h(q) \quad (3)$$

Here, x_1 and x_2 are the relative composition ratios of the spherical and cylindrical micelles, respectively. The solid lines on the data points are resultant best fit curves, showing the Eq 3 well reproduces the experiments. Figure 2b plots X and g (the paracrystalline distortion factor) against N/P, showing that the structural order was increased upon addition of DNA. Here, X is the cylinder composition determined by $1\text{-}x_1/x_{1,micelle}$, $x_{1,micelle}$ is the relative composition of the spherical micelles before addition of DNA. The profiles at 3.3 > N/P in Figure 2a showed that further addition of DNA to the cylinders induced transition to a lamella structure. This transition was completed around N/P = 1.0 and further addition of DNA did not induce any changes. N/P = 1.0 means that one-to-one interaction is attained and thus all of the BA molecules are bound to DNA. Since DNA is included in the hydrophobic domain, DLPC molecules are seemed to face outside of the lamella as depicted in Figure 5.

3-2: Structural Transition at Point B

Figure 4a presents N/P dependence of SAXS at the composition of B (the highest transfection efficiency). Before adding DNA, the profile had the peaks that can be characterized to a hexagonally packed cylinder overlapping a convexed base line. This convex curve may be ascribed to an isolated scattering object (or a form factor). With adding DNA, the first diffraction peak becomes sharper and more intense and other higher-order peaks become obvious, while the convex curve becomes less obvious. These features can be interpreted by that the addition of DNA did not induce structural transitions such as the composition A, but to enhance the hexagonal ordering as well as reducing the amount of the isolated scattering object. The data was fit by assuming that the solution consists of both hexagonally-packed and isolated cylinders with the same structural parameters:

$$I(q) = yP_C(q) + (1-y)P_C(q)S_h(q) \quad (4)$$

The solid lines in the figure 4a represent the best fit and the parameters are summarized in Table 1. Figure 4b plots obtained g and $1\text{-}y$ values against N/P. As presented in Figure 4a, all of the profiles could be fit with the same cylinder with the inner structure depicted in Figure 5b, confirming that the addition of DNA did not induce structural transition but enhanced the packing of the cylinders.

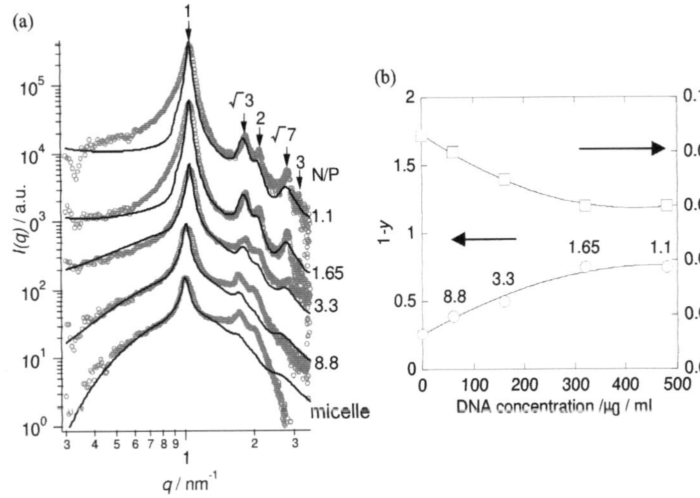

Figure 4. (a) N/P dependence of SAXS profile at the composition B (BA:DOPE:DLPC=1:2:1). The micelle concentrations are 3.2mM at N/P=1.65 - 8.8 or 2.13mM at N/P=1.1. (b) N/P dependence of g (□) and $1\text{-}y$ (○).

4. Concluding Remarks

Figure 5 depicted two models to explain N/P dependence of SAXS for the A and B points. At the point A, SAXS showed that drastic changes took place upon addition of DNA, from a spherical micelle to a hexagonally packed cylinder, then to lamella. During this process, DNA is considered to be incorporated into the hydrophobic domain. At the point B, the preceding cylinders are gathered by DNA and they are intercalated with each other, without undergoing major structural changes; presumably anionic DNA might neutralize and thus joint the cationic cylinders. With adding DNA, the inter-cylinder distance decreased from 7.16 nm to 6.85 nm, confirming the DNA's role of an adhesion. Comparison of these two compositions suggests that difference in the DNA location in the complex is related to the transfection efficiency.

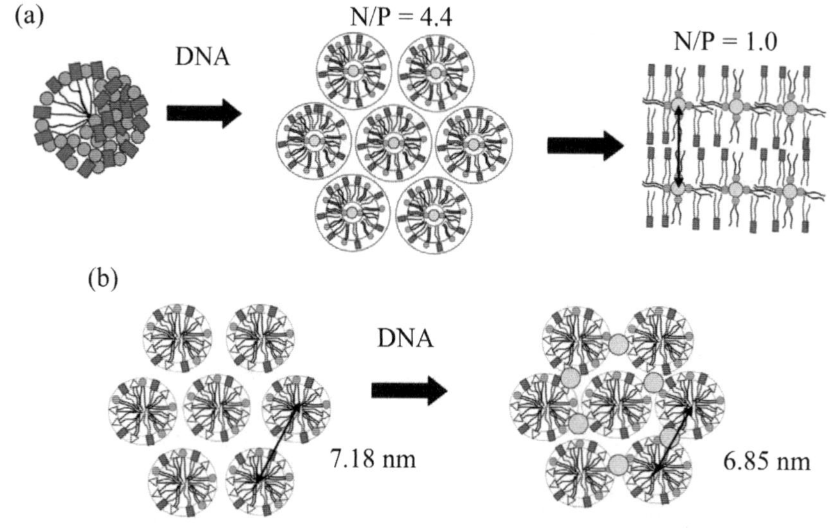

Figure 5. Schematic drawing of the morphological changes upon mixing (a) A-micelle and pDNA and (b) B-micelle and pDNA.

Acknowledgements

This work is financially supported by the JST CREST program and the Ministry of Education, Science, Grant-in-Aid for 20550193 (Akiba). All SAXS measurements were carried out at SPring-8 40B2 (2008B0012 for Sakurai and 2008B1781 for Akiba). Our thanks extend to Prof. Masaki Takada and Dr. Sono Sasaki for fruitful discussion on SAXS analysis and instrumentation.

References

[1] Felgner P L and Rhodes G 1991 *Nature* **349** 351
[2] Radler J O, Koltover I, Salditt T and Safinya C R 1997 *Science* **275** 810
[3] Lin A J, Slack N L, Ahmad A, Koltover I, George C X, Samuel C E and Safinya C R 2000 *Journal of Drug Targeting* **8** 13
[4] Huang L, Hung M C and Wagner E 1999 *Nonviral Vectors for Gene Therapy* (Academic Press London UK) p 91
[5] Sakurai K, Kusuki S, Hamada E and Takeda Y, JPA 2008-143844.
[6] Ewert K K, Samuel C E and Safinya C R 2008 *DNA Interactions with Polymers and Surfactants*
[7] Sakurai K, Synchrotron Small-Angle X-ray Scattering from Ultra-Dilute Solutions of pDNA/Cationic Lipid Complexes, X-ray and Neutron Techniques for Nano-Structural Research, SPring-8, Hyogo, Japan, 2008.
[8] Guinier A and Fournet G 1955 *Small-Angle Scattering of X-rays* (Inc.: New York)
[9] Hashimoto T, Kawamura T, Harada M and Tanaka H 1994 *Macromolecules* **27** 3063
[10] Kawaguchi T, Hamanaka T, Kito Y and Machida H 1991 *J. Phys. Chem.* **95 (9)** 3837
[11] Berne B J and Pecora R 2000 *Dynamic Light Scattering with Applications to Chemistry, Biology, and Physics* (Inc.:Dover Publications)

Journal of Physics: Conference Series **184** (2009) 012009 doi:10.1088/1742-6596/184/1/012009

Molecular Aggregation States of Poly{2-(perfluorooctyl)ethyl acrylate} Polymer Brush Thin Film Analyzed by Grazing Incidence X-ray Diffraction

H. Yamaguchi[1], K. Honda[1], M. Kobayashi[2], M. Morita[3], H. Masunaga[4], O. Sakata[5], S. Sasaki[4,5], M. Takata[4,5,6], and A. Takahara[1,2,5]

[1] Graduate School of Engineering, Kyushu University, Nishi-ku, Fukuoka 819-0395, Japan. [2] Institute for Materials Chemistry and Engineering, Kyushu University, 744 Motooka, Nishi-ku, Fukuoka 819-0395, Japan. [3]Fundamental Research Department, Chemical Division, Daikin Industries, Ltd.,1-1 Nishi Hitotsuya, Settsu-shi, Osaka 566-8585, Japan. [4] Japan Synchrotron Research Institute, Mikazuki Sayo, Hyogo 671-5198, Japan. [5] The RIKEN Harima Institute, Mikazuki Sayo, Hyogo 671-5198, Japan. [6]Department of Advanced Materials Science, School of Frontier Science, The University of Tokyo, Kashiwa 277-8561, Japan

E-mail: takahara@cstf.kyushu-u.ac.jp

Abstract. Fluoropolymer brush with crystalline side chains was prepared by surface-initiated atom transfer radical polymerization of 2-(perfluorooctyl)ethyl acrylate (FA-C_8) from a flat silicon substrate. The crystallization and the molecular aggregation structures of polymer side chain at the outermost surface and internal region in the brush film were characterized by grazing incidence X-ray diffraction (GIXD) measurement using two different incident angles of X-ray. At the air interface of PFA-C_8 brush film, the rod-like R_f group was oriented perpendicular to the surface forming a hexagonal packing structure to reduce surface energy. In contrast, the oriented R_f groups parallel to the substrate coexisted at the internal region in the brush. This unique depth dependence of crystalline state of the fluoropolymer brush was observed by surface-sensitive GIXD measurement.

1. Introduction

Surface-initiated polymerization from a solid surface gives a tethered polymer with high graft density and controlled molecular weight, a so-called "polymer brush" [1]. Recent development of "grafting-from" method supported by living polymerization made it possible to prepare the highly dense brush [2]. In general, the graft density of the polymer chains is so high that the polymer chains are forced to stretch along the direction normal to the graft surface [3]. Therefore, the molecular mobility and chain conformation of brush are restricted to form a unique molecular aggregation structure [4].

The long perfluoroalkyl (R_f) chains of poly(fluoroalkyl acrylate) also favor to form an aggregated structure aligning perpendicular direction to the substrate to reduce the surface energy [5]. It is known that the molecular aggregation state of the R_f groups strongly affects the surface wettablity and water repellency of the polymer films [6]. Previously, the authors prepared poly{2-(perfluorooctyl)ethyl acrylate} (PFA-C_8) brushes with thicknesses of 4 and 11 nm on silicon wafer by surface-initiated atom transfer radical polymerization (ATRP) to analyze the molecular aggregation structure of R_f groups by wide-angle X-ray diffraction, X-ray reflectivity (XR), and grazing incident X-ray diffraction (GIXD) measurements [7]. The diffraction pattern attributed to the crystalline and ordered structure of the R_f groups, including a lamellar structure, was observed from the thicker brush film, while the thinner PFA-C_8 film showed only an amorphous hallow without any sharp diffraction peaks. The R_f groups located near to the brush/substrate interface could not diffuse enough to form an ordered packing structure due to the immobilization of main chain.

In this paper, we tried to characterize the depth profile of the molecular aggregation structure of PFA-C_8 brush with thickness of 41 nm by GIXD in order to elucidate the relationship between the brush thickness and the molecular aggregation state of crystalline side chains to show the structure along with the distance from the outermost brush surface to the substrate interface.

2. Experimental

2.1. Materials. Copper (I) bromide (CuBr, Wako Pure Chemicals, 98%) was purified by washing with acetic acid and ethanol and was dried under vacuum. Ethyl 2-bromoisobutylate (EB) (TCI, 99%), 2,2'-bipyridyl (Wako, 98 %) and 4,4'-dinonyl-2,2'-bipyridyl (C_9Bpy, Aldrich, 98 %) were used as received. FA-C_8 monomer (Daikin Industry Ltd.) was purified by distillation from calcium hydride under reduced pressure. The silicon (111) wafers (diameter = 1 inch, thickness = 3 mm) were cleaned by washing with piranha solution and by exposure to vacuum ultraviolet-ray (VUV, λ = 172 nm) for 10 min under reduced pressure. Surface initiator, (2-bromo-2-methyl) propionyloxyhexyltriethoxysilane (BHE) [8,9] was immobilized on the silicon wafers by the chemical vapor adsorption method to form a monolayer as reported previously [10]

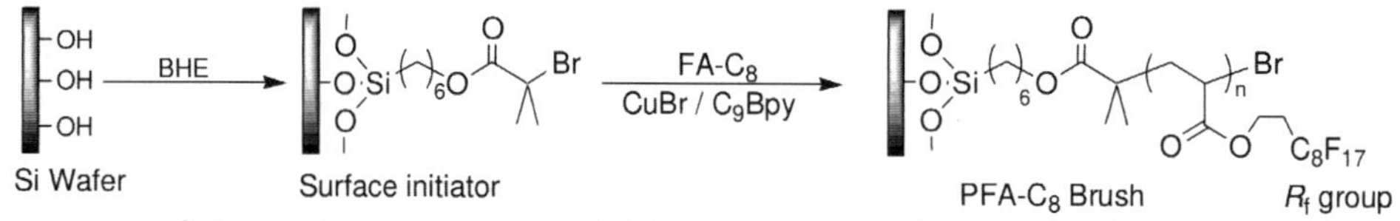

Scheme 1. Preparations of PFA-C_8 brush by surface-initiated ATRP.

2.2. General procedure for surface-initiated ATRP. Preparation procedure of PFA-C_8 brush was described elsewhere [7]. As shown in Scheme 1, surface-initiated ATRP of FA-C_8 from the initiator-immobilized silicon wafers was carried out in the presence of CuBr, C_9Bpy, and EB at 383 K for 72 h under argon. The resulting silicon wafer was washed with AK-225 (fluoro organic solvent, Asahi Glass Co.) using a Soxhlet apparatus for 8 h to remove the free polymer absorbed on their surface, and were dried under reduced pressure. Sample was annealed at 373 K for 1 h and gradually cooled to room temperature at 10 K/min.

2.3. Measurements. XR and GIXD measurement were carried out on a BL13XU beam line of SPring-8 (Japan Synchrotron Radiation Research Institute, Hyogo, Japan) using an incident X-ray with a wavelength λ of 0.100 nm [11]. The scattering vector, q, in specular reflectivity is defined by $q = (4\pi/\lambda)\ sin\theta$. The incident and reflection angles of XR were kept at same magnitude for specular condition. Model calculation and data analysis for the XR were carried out using a software, XRR, a program provided by Rigaku Corp. GIXD data were obtained from the surface and deeper regions at α_i of 0.08 and 0.16, respectively [12,13]. Diffractions from the sample were detected in the in-plane and out-of-plane directions. The information about the structure perpendicular to the surface is obtained from in-plane geometry because scattering vector (q_{xy}) is parallel to the surface. On the other hand, the detected profiles reflect information on crystalline states parallel to the film surface in the out-of-plane geometry. XR and GIXD observations were carried out at room temperature.

3. Results and Discussion

3.1. X-ray reflectivity of PFA-C_8 brush. Figure 1(a) shows a reflectivity profile of PFA-C_8 brush film, and corresponding fit which was calculated by the density distribution of the thin film as shown in

Figure 1(b). The thickness of the brush film can be roughly estimated from the intervals of the Kiessig fringes in the XR curve. The bulk density of silicon and SiO_2 are 2.33 and 2.65 g cm^{-3}, respectively. As can be seen from Figure 1(b), the bulk density inside the brush layer near the tethered point at the silicon substrate was relatively low and the density gradually increased along with the distance from the substrate. The outermost area facing air revealed the highest density of 2.3 g cm^{-3}, which is close value to the density of poly(tetrafluoroethylene) (PTFE). Thus, the density distribution of the PFA-C_8 brush thin film is not homogeneous. In other words, molecular chains seem to form ordered state at the outermost surface of PFA-C_8 thin films.

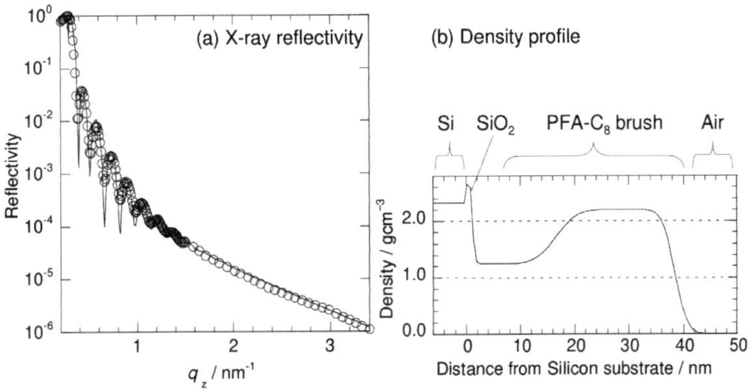

Figure 1. XR curve of PFA-C_8 brush (circle plot) and the corresponding (solid line) (a). The assumed density profile is shown in right side (b).

3.2. GIXD measurement of PFA-C_8 brush. Figure 2(a) shows the in-plane GIXD profiles of PFA-C_8 brush film. Sharp and strong peaks were appeared for PFA-C_8 brush thin films at 12.6 nm^{-1}. Hence, the side chains of the PFA-C_8 are oriented perpendicular to the surface. The *d*-spacing calculated from the peak position was ca. 0.50 nm, which was close to the intermolecular distance between R_f chains of PTFE hexagonal (d = 0.49 nm) [14]. Therefore, it was suggested that the rigid rod-like R_f groups underwent hexagonal packing at the outermost surface of PFA-C_8 brush film. Figure 2(b) represents the out-of-plane GIXD profile of PFA-C_8 brush film. The strong diffraction peaks were observed at q_z = 2.2, 3.9, and 5.8 nm^{-1} correspond to the *d*-spacings of 2.84, 1.56, and 1.05 nm, respectively. These peaks are assignable to the lamellar structure of R_f group. These Bragg diffractions indicate that the (001) crystallographic planes are oriented almost parallel to the substrate.

Interestingly, a broader diffraction peak at q_z = 12.5 nm^{-1} was observed from out-of-plane measurement with higher incident angle $\alpha_i = 0.16°$, while no peak appeared at the same q_z range in the diffraction pattern obtained by lower α_i (0.08°). This result suggested that the hexagonal packing structure of R_f groups at the deeper region in the brush thin film was oriented parallel to the substrate. In addition, in-plane diffraction pattern in the case of $\alpha_i = 0.16°$ showed a peak at q_{xy} = 3.9 and 5.9 nm^{-1} attributed to lamellar structure.

Considering these X-ray diffraction patterns, the R_f aggregation states seem to be different between the outermost surface and internal region in the brush film, as shown in Figure 2(c). At the air interface of PFA-C_8 brush, the rod-like R_f groups oriented perpendicular to the substrate forming a hexagonal packing structure, which contribute to the reduction of surface energy. In contrast, the oriented R_f groups parallel to the substrate coexisted at the internal region in the brush. We suppose that the densely immobilized polymer backbones were forced to stretch at perpendicular direction to the silicon wafer, resulting in a parallel orientation of R_f side chains to the substrate. However, the brush/substrate interface region of brush seems to contain an amorphous structure of R_f groups because broader diffraction peak was observed by GIXD at higher α_i.

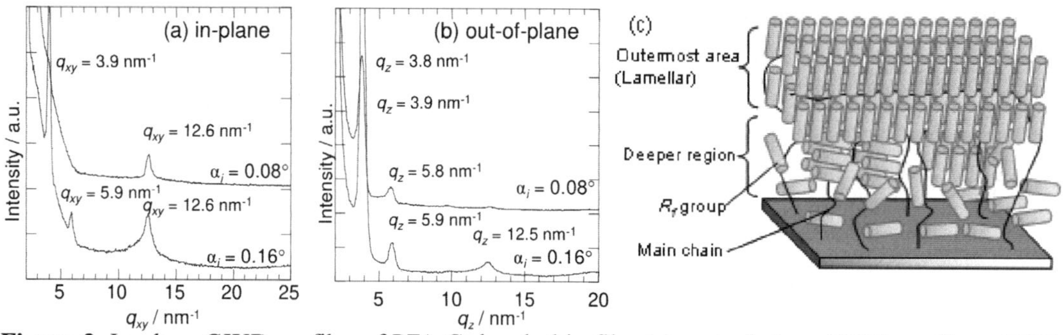

Figure 2. In-plane GIXD profiles of PFA-C_8 brush thin film (a), out-of-plane GIXD profiles of PFA-C_8 brush thin film(b), and schematic image of the molecular aggregation structure of R_f groups in PFA-C_8 brush (c).

4. Conclusion

Semicrystalline polymer brushes having R_f groups was prepared by surface-initiated ATRP on silicon substrate. The surface- and bulk-sensitive GIXD measurements of polymer brushes film showed the different aggregation state of the R_f groups at the air/brush and the brush/substrate interfaces. On the outermost surface of PFA-C_8 brush film, R_f groups formed a hexagonal packing structure due to the low surface free energy, and were oriented almost perpendicular to the substrate to afford a lamellar structure. On the other hand, the R_f groups were partially oriented parallel to the substrate at the deeper layer in the brush. The anchoring effect of polymer backbone restricted the available chain conformation and molecular mobility of R_f groups at the brush/substrate interface to form different molecular aggregation structure from that of the air/brush interface.

Acknowledgements

The synchrotron X-ray diffraction experiments were performed at the BL13XU in the SPring-8 with the approval of JASRI (Proposals No.2007A1429 & No. 2007B1530).

References

[1] Advincula R C, Brittain W J, Caster K C and Ruhe J 2004 *Polymer Brushes: Synthesis, Characterization, Applications* (Weinheim, Wiley VCH).
[2] Jordan R 2006 *Surface-Initiated Polymerization I* (Garching: Springer).
[3] Yamamoto S, Ejaz M, Tsujii Y, Matsumoto M and Fukuda T 2000 *Macromolecules* **33** 5602.
[4] Uekusa T, Nagano S and Seki T 2007 *Langmuir* **23** 4642.
[5] Honda K, Morita M, Otsuka H and Takahara A 2005 *Macromolecules* **38** 5699.
[6] Honda K, Yakabe H, Koga T, Sasaki S, Sakata O, Otsuka H and Takahara A 2005 *Chem. Lett.* **34** 1024.
[7] Yamaguchi H, Honda K, Kobayashi M, Morita M, Masunaga H, Sakata O, Sasaki S and Takahara A 2008 *Polym. J.* **40** 854.
[8] Ohno K, Morinaga T, Koh K, Tsujii Y and Fukuda T 2005 *Macromolecules* **38** 2137.
[9] Kobayashi M and Takahara A 2005 *Chem. Lett.* **34** 1582.
[10] Kobayashi M, Terayama Y, Hosaka N, Kaido M, Suzuki A, Yamada N, Torikai N, Ishihara K and Takahara A 2007 *Soft Matter* **3** 740.
[11] Sakata O, Furukawa Y, Goto S, Mochizuki T, Uruga T, Takeshita K, Ohashi H, Ohata T, Matsushita T, Takahashi S, Tajiri H, Ishikawa T, Nakamura M, Ito M, Sumitani K, Takahashi T, Shimura T, Saito A and Takahashi M 2003 *Surf. Rev. Lett.* **10** 543.
[12] Russell T P 1990 *Mater. Sci. Rep.* **5** 171.
[13] Yakabe H, Tanaka K, Nagamura T, Sasaki S, Sakata O, Takahara A and Kajiyama T 2005 *Polym. Bull.* **53** 213.
[14] Bunn C W and Howells E R 1954 *Nature* **174** 549.

Characterization of Swollen Structure of High-density Polyelectrolyte Brushes in Salt Solution by Neutron Reflectivity

Motoyasu Kobayashi[1,2], Yuki Terayama[3], Masahiro Hino[4], Kazuhiko Ishihara[5], Atsushi Takahara[1,2,3]

[1] Institute for Materials Chemistry and Engineering, Kyushu University, 744 Motooka, Nishi-ku, Fukuoka 819-0395, Japan. [2] JST/ERATO, [3] Graduate School of Engineering, Kyushu University, [4] Reactor Research Institute, Kyoto University, [5] Graduate School of Engineering, The University of Tokyo

E-mail: takahara@cstf.kyushu-u.ac.jp

Abstract. Zwitterionic and cationic polyelectrolyte brushes on quartz substrate were prepared by surface-initiated atom transfer radical polymerization of 2-(methacryloyloxy)ethyl phosphorylcholine (MPC) and 2-(methacryloyloxy)ethyltrimethylammonium chloride (METAC), respectively. The effects of ionic strength on brush structure and surface properties of densely grafted polyelectrolyte brushes were analysed by neutron reflectivity (NR) measurements. NR at poly(METAC)/D_2O and poly(MPC)/D_2O interface revealed that the grafted polymer chains were fairly extended from the substrate surface, while the thickness reduction of poly(METAC) brush was observed in 5.6 M NaCl/D_2O solution due to the screening of the repulsive interaction between polycations by hydrated salt ions. Interestingly, no structural change was observed in poly(MPC) brush even in a salt solution probably due to the unique interaction properties of phosphorylcholine units.

1. Introduction

The dimensions and hydrodynamic radius of the isolated polyelectrolytes bearing positive or negative charges in an aqueous solution strongly depend on the ionic strength of the solution [1,2]. It is generally accepted that polyanion or polycation chains in aqueous solution with low ionic strength form expanded chain structures due to the intramolecular repulsive interactions. Electrostatic repulsive interactions also enhance intermolecular repulsive interactions, resulting in a large second virial coefficient. In contrast, the polyelectrolyte dissolved in higher ionic strength in solution behaves like an electrically neutral polymer to give smaller dimension because the electric interactions are screened by salt ions. However, surface-grafted polymer chains are expected to exhibit different behaviors from those of the isolated polymers [3].

The dimension of nonionic polymer chains immobilized on solid surface largely depends on the graft density. Polymer chains sparsely immobilized on a flat surface form a coil structure such as a mushroom, whereas densely immobilized polymer chains on a flat surface in a good solvent stretch in a perpendicular direction against the substrate surface because of high osmotic pressure arising from the excluded volume effect by densely-grafted neighboring chains [4]. In the case of polyelectrolyte brush, repulsive electrostatic interaction would also afford a stretched chain conformation. Then, we are wondering whether screening of electrostatic interaction by salt ions could take place for the densely-grafted polyelectrolyte chains to change the extended chain structure or not. The behavior of polyelectrolyte brushes in salt solution is important for applications in medical materials, as these materials are in contact with blood and other body fluids containing salts. Herein, we describe the dependence of swelling brush structure on the ionic strength analyzed by neutron reflectivity (NR) of zwitterionic type polyelectrolyte, as well as cationic polyelectrolyte brushes prepared by surface-initiated controlled polymerization of 2-(methacryloyloxy)ethyl phosphorylcholine (MPC) and 2-(methacryloyloxy)ethyltrimethylammonium chloride (METAC), respectively.

2. Experimental

Quartz plates (65 × 65 × 10 mm^3) were cleaned by washing with piranha solution at 373 K for 1 h and subsequent exposure to vacuum ultraviolet-ray (λ = 172 nm) for 5 min under reduced pressure (30 Pa). The surface initiator, (2-bromo-2-methyl)propionyloxyhexyltriethoxysilane, was immobilized on the substrate by chemical vapour adsorption method.[5] Poly(MPC) brush was prepared by surface-initiated atom transfer radical polymerization in methanol using CuBr and 4,4'-dimethyl-2,2'-bipyridyl at 303 K for 12 h.[6] For METAC polymerization, 10% isopropanol aqueous solution was used as a solvent [7]. The number-average molecular weights (M_n) and molecular weight distribution of the unbound polymers were determined by size-exclusion chromatography recorded on a JASCO instrument equipped with a JASCO 2031plus RI detector and TSK-gel column Super H6000, H4000, and H2500 using 0.01 M LiBr aqueous solution for poly(MPC) and acetonitrile/water (1/1, v/v) solution with 0.5 M of acetic acid and 0.2 M sodium nitrate for poly(METAC) as the eluents at a rate of 0.5 mL/min at 313 K. The thickness of the polymer brushes on the silicon substrate were determined using an imaging ellipsometer (Nippon Laser & Electronics Lab.) equipped with a YAG laser (532.8 nm). The polarizer angle was fixed at 50°, and a refractive index of 1.48 was assumed for calculations of the film thickness. NR measurements were carried out using a multilayer interferometer for neutrons (MINE) in JRR-3 at TOKAI, using wavelength λ = 0.88 nm. A neutron beam was irradiated from quartz to the interface between D_2O and the poly(MPC) brush on quartz glass. The incident slits were adjusted to maintain a 55 mm footprint size on the sample surface. The NR profiles were analyzed by fitting calculated reflectivity from model scattering length density profiles to the data, using Parratt32 software.

3. Results and Discussion.

Figure 1 shows the NR curves of D_2O/poly(METAC) brush interface and the corresponding neutron scattering length density (SLD) profile normal to the interface. A critical angle was clearly observed at q = 0.11 nm^{-1} attributed to the difference in SLD between quartz and deuterium oxide. The neutron SLD profiles of the poly(METAC) brush in D_2O formed in a smooth upward curve from 5.2 × 10^{-4} to 6.36 × 10^{-4} nm^{-2} along with the distance from the substrate. The gradient profile indicated that the polymer chains in D_2O were stretched up to ca. 80 nm, which is larger than the brush thickness under the air (*humidity* = 50%) estimated to be 46 nm by ellipsometer. Therefore, the polymer chains immersed in D_2O were considerably extended to perpendicular direction against the substrate. High osmotic pressure caused by high concentration of quarternary ammonium ions in the brush layer would have made an extended chain conformation. The gradual change in SLD was appeared by the molecular weight distribution of polymer brush. Large polydispersity index of brush, M_w/M_n = 1.5, afforded a thick boundary zone to make unclear fringes in reflectivity curve.

In general, an increase in ionic strength leads to the reduction of the electrostatic repulsion. The polyelectrolyte chains in a salt solution may undergo a transition toward a more coiled state and show a volume reduction. However, quite similar reflective curve was observed from the interface of the poly(METAC) brush in the 1.0 M NaCl/D_2O solution, indicating that the extended structure of the swelling brush chains did not change, even in a salt solution. It is a quite contrast to the fact that the unbound polyelectrolyte would shrink in salt solution. We estimated the swollen brush thickness by volume fraction (φ) curve of METAC segments as a function of the distance from substrate (z) and the corresponding parabolic-type profile assuming a brush layer height (h^*). A pure parabola equation of $\varphi = \varphi_0\{(1-z/h^*)^2\}$ is theoretically predicted by Cates *et al.* for polymer brush in the limit of strong stretching.[8] Actual volume fraction curve as shown in Figure 2 requires the parabolic function and an additional exponential tail at larger z region to obtain better fit due to a roughness at the brush/solution interface. We regard h^* as thickness of the swollen brush. Let us assume the swollen poly(METAC) thickness in 1.0 M NaCl solution was 83 nm, the M_n was 86,000, and grafting density was 0.36 chains/nm^2, the apparent concentration of METAC monomers in the brush layer can be estimated to be 3.1 M, which is much higher than salt concentration in the external solution. Therefore, hydrated salt ion could not diffuse into a high-density polymer brush layer due to the high local charge density, and then polyelectrolyte brush could keep an extended chain structure in solution. Similar results have been reported by Matsuoka [9, 10], Kurihara [11], and Tran.[12] Huck *et al.* reported the swelling behavior of the poly(METAC) brush in solution observed by AFM.[13] They suggested that thickness of brush decreased at much lower external salt concentrations than the concentration of chemically charged monomers of the chains. Unfortunately, our NR profiles could not detect the apparent change in the swollen structure of brush between at 0 ~ 1 M NaCl/D_2O solution probably due to the large roughness and nonuniform thickness of the brush along the neutron-irradiated footprint as long as 50 mm.

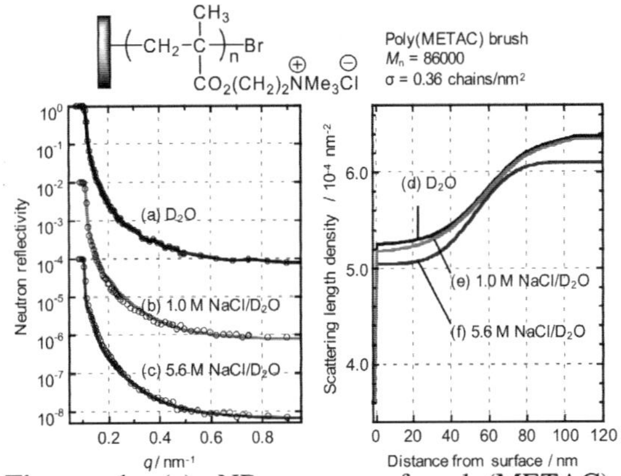

Figure 1. (a) NR curves of poly(METAC) brush/D_2O, (b) poly(METAC) brush/ 1.0 M NaCl in D_2O, (c) poly(METAC) brush/5.6 M NaCl in D_2O, and (d)-(f) their corresponding neutron SLD profiles along with the distance from quartz surface, respectively. Scattering vector $q = 4\pi \sin\theta / \lambda$ at $\lambda = 0.88$ nm.

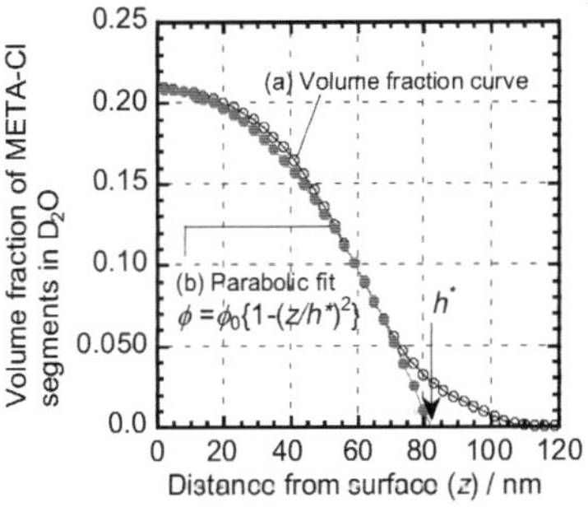

Figure 2. (a: open circle) Volume fraction distribution (φ) for poly(METAC) brush in 1.0 M NaCl/D_2O calculated by Figure 1(e), and (b: filled circle) the corresponding fit based on a Parabolic function using $\varphi = \varphi_0\{(1-z/h^*)^2\}$.

When the poly(METAC) brush was immersed in the 5.6 M NaCl/D_2O solution, the reduction in roughness and thickness of swelling brush layer was observed, as shown in Figure 1(f). The brush layer height h^* was 69 nm. The hydrated salt ions in the adjacent solution with higher ionic strength than the brush layer must have screened the repulsive interaction between quarternary ammonium groups of the brush chain to moderate highly extended chain conformation. The graft area occupied with a tethered chain was 2.7 nm^2 given by graft density. Considering the molecular volume per monomer unit v_0 (= 0.28 nm^3) estimated from bulk density of METAC and the chain contour length per monomer unit l_0 (l_0= 0.25 nm for vinyl monomers), the cross-sectional area per monomer unit $a^2 = v_0/l_0$ is 1.1 nm^2. Therefore, the distance between the tethered chains on the substrate surface is around 1.0 nm, which is larger than size of hydrated ions. Apparent hydration radius of equivalent hydrated sphere of Na^+ and Cl^- are reported to be 0.18 and 0.20 nm.[14] Once the concentration of added ions in solution reaches the concentration of the free counterions inside the brush, hydrated ions can enter the polyelectrolyte brush layer to screen the the electric interactions forming relatively smaller dimension and thickness. The solvent quality of saturated NaCl aqueous solution for poly(METAC) has not been well-studied yet, although poly(METAC) quickly dissolves in 5.6 M NaCl aqueous solution. The second virial constant of poly(METAC) in NaCl solution should be given by light scattering measurement.

Independency on salt concentration was also observed between the swelling structure of poly(MPC) facing a pure D_2O and a NaCl/ D_2O solution, as shown in Figure 3(a) and (b). Swollen thickness h^*

was estimated to be around 90 nm. As mentioned before, higher ionic strength in densely grafted poly(MPC) layer than that in outer solution might be one of the reason for a negligible change in swollen brush structure. In this case, however, additional mechanism based on a specific property of MPC should be also contributed. We previously carried out dynamic light scattering measurement of unbound free poly(MPC) dissolved in an aqueous solution to find out that the dimension of poly(MPC) was independent on the ionic strength of the salt solution.[15] Poly(MPC) can be regard as a quite unique zwitter-type polyelectrolyte of which chain structure in a aqueous solution hardly changed by salt effect, although its reason has not been cleared, yet.

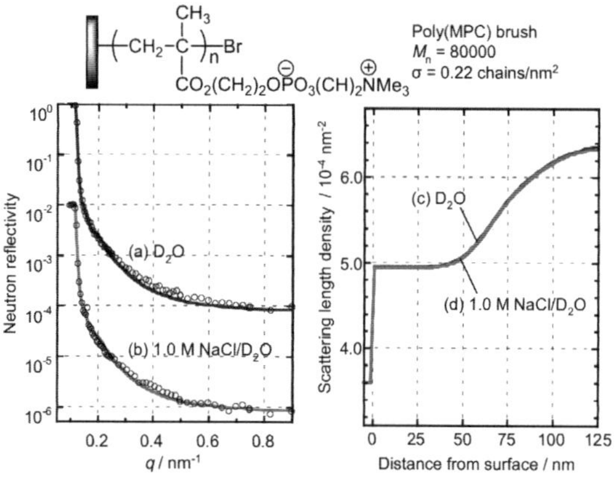

Figure 3. (a) NR curves of poly(MPC) brush/D_2O, (b) poly(MPC) brush/ 1.0 M NaCl in D_2O, and (c)-(d) their corresponding neutron SLD profiles along with the distance from quartz surface, respectively. Scattering vector $q = 4\pi \sin\theta / \lambda$, at $\lambda = 0.88$ nm.

4. Conclusion

NR of the swollen poly(METAC) and poly(MPC) brushes revealed that the extended chain structure formed in pure water as well as in aqueous NaCl solution with low concentration (1.0 M). We suppose that the hydrated ions of sodium and chloride in the external solution could not diffuse into the swollen brush layer due to the high osmotic pressure and high charge density of the densely packed polyelectrolyte chains. At the higher salt concentration (5.6 M), the screening of repulsive interaction took place to reduce the thickness of poly(METAC) brush. Negligible dependency of hydrated poly(MPC) behavior on salt concentration is supposed be based on the unique interaction of phosphorylchorine and water molecule, however, further investigation using various derivatives of MPC and analysis at higher salt concentration will be required to discuss the details.

Acknowledgments
The present work was supported by a Grant-in-Aid for Young Scientist (B) (19750098) from the Ministry of Education, Culture, Science, Sports and Technology of Japan.

References

[1] Takahashi A, Kato N and Nagasawa M 1970 *J. Phys. Chem.* **74** 944.
[2] Nagasawa M and Eguchi Y 1976 *J. Phys. Chem.* **71** 880.
[3] Advincula R C, Brittain W J, Caster K C and Rühe J 2004 *Polymer Brushes* (Wiley-VCH, Weinheim) p.331-370.
[4] Tsujii Y, Ohno K, Yamamoto S, Goto A and Fukuda T 2006 *Adv. Polym. Sci.* **197** 1.
[5] Kobayashi M and Takahara A 2005 *Chem. Lett.* **34** 1582.
[6] Kobayashi M, Terayama Y, Hosaka N, Kaido M, Suzuki A, Yamada N, Torikai N, Ishihara K and Takahara A 2007 *Soft Matter* **3** 740.
[7] Li Y, Armes S P, Jin X and Zhu S 2003 *Macromolecules* **36** 8268.
[8] Milner S T, Witten T A and Cates M E 1988 *Macromolecules* **21** 2610
[9] Kaewsaiha P, Matsumoto K, and Matsuoka H 2004 *Langmuir* **20** 6754.
[10] Kaewsaiha P, Matsumoto K, and Matsuoka H 2007 *Langmuir* **23** 7065.
[11] Sanjuan S, Perrin P, Pantoustier N, and Tran T 2007 *Langmuir* **23** 5769.
[12] Hayashi S, Abe T, Higashi N, Niwa M, Kurihara K 2002 *Langmuir* **18** 3932.
[13] Moya S E, Azzaroni O, Kelby T. Donath E and Huck W T S 2007 *J. Phys Chem* **111** 7034.
[14] Kiriukhin M Y and Collins K D 2002 *Biophys Chem* **99** 155.
[15] Matsuda Y, Kobayashi M, Annaka M, Ishihara K and Takahara A 2008 *Langmuir* **24** 8772.

Influence of Salt Concentration on Swelling States of Poly(sulfobetaine) Brush at Aqueous Solution Interface

Yuki Terayama[1], Moriya Kikuchi[2,3], Motoyasu Kobayashi[2,3], Masahiro Hino[4], Atsushi Takahara[1,2,3]

[1]Graduate School of Engineering, Kyushu University, [2]Institute for Materials Chemistry and Engineering, Kyushu University, [3]JST/ERATO, 744 Motooka, Nishi-ku, Fukuoka 819-0395, Japan. [4]Reactor Research Institute, Kyoto University, Kumatori-cho, Sennan-gun, Osaka 590-0494, Japan

E-mail: takahara@cstf.kyushu-u.ac.jp

Abstract. Swelling states of poly(3-dimethyl(methacryloyloxyethyl)ammonium propane sulfonate) (DMAPS) brush in aqueous solutions with different salt concentrations were investigated by atomic force microscopy (AFM) and neutron reflectivity measurement. The thickness of swollen poly(DMAPS) brush evaluated by AFM was increased from 212 nm up to 352 nm with an increase in salt concentration from 0 to 0.5 M. Poly(DMAPS) brush chains formed shrunk structure in a pure water due to the attractive electrostatic inter- and intra-chain interaction of sulfobetaine groups, while the brush chains in NaCl aqueous solution were stretched up because the hydrated salt ions screened the attractive interaction.

1. Introduction

Polysulfobetaine, such as poly(3-dimethyl(methacryloyloxyethyl)ammonium propane sulfonate) poly(DMAPS), is one of the zwitterionic polymers, which has attracted much attention due to its unique solution properties and applications. For instance, poly(DMAPS) cannot be dissolved in pure water due to the strong attractive electrostatic interaction between ammonium cations and sulfonyl anions of the side chains. Salt ions added into the poly(DMAPS) aqueous solution screen the inter- or intramolecular interaction among side chains to give a greater chain expansion [1], higher intrinsic viscosity [2], and lower upper critical solution temperature (UCST) [3]. Not only solution properties but also surface-tethered poly(DMAPS) on a substrates was investigated by several groups. Kato *et. al.* estimated the thickness of adsorbed poly(DMAPS) layer on silica surfaces as function of salt concentration [4]. Very low non-specific protein adsorption was observed on the surface of poly(DMAPS)-immobilized substrate prepared by surface-initiated atom transfer radical polymerization (ATRP) [5]. The wettability control of the substrate surface has been also achieved by using poly(DMAPS) brush with various ranges of graft density [6]. They described that the wettability of poly(DMAPS) brush switched from hydrophilic to hydrophobic at UCST. The switching temperature became higher by increasing the graft density. In general, the high-density polymer brush chains are extended to perpendicular direction from a substrate in good solvent due to the osmotic pressure by excluded volume effect [7]. In the case of polyelectrolyte brush, the repulsive or attractive interaction among the ionic groups of polymer chains also affects the swelling structure of the brush in the solution [8]. In particular, poly(DMAPS) brush chains would form a relatively extended conformation in a salt water, while the chains would shrink in pure water attributed to the strong attractive interaction among side chains. In this report, the dependence of ionic strength on swelling behavior of poly(DMAPS) was investigated by atomic force microscopy (AFM) and neutron reflectivity (NR) measurement.

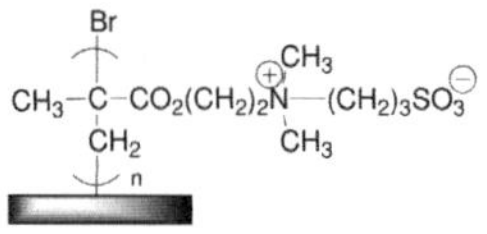

Figure 1. Poly (DMAPS) brush.

2. Experimental

2.1. Materials. The surface initiator, (2-bromo-2-methyl)propionyloxyhexyltriethoxysilane (BHE) [9,10] and DMAPS monomer [11] were synthesized using previous reported procedures. Silicon wafer and quartz plates ($65 \times 65 \times 10$ mm^3) were washed with piranha solution at 373 K for 1 h and subsequently exposed to the vacuum ultraviolet-ray for 15 min under 30 Pa. The BHE monolayer was immobilized on the substrates using the chemical vapour adsorption (CVA) method [12].

2.2. Surface-initiated ATRP. A few sheets of BHE-immobilized silicon wafers and quartz plate, CuBr (0.088 mmol), and 2,2'-dipyridyl (0.168 mmol) were introduced into a glass tube equipped with a stopcock under argon atmosphere. A methanol/water (1/1 = v/v) solution of DMAPS (215 mmol) was added to the catalyst in glass tube, and was degassed three times by repeated freeze-pump-thaw cycles. Polymerization reaction was carried out at 303 K for 12 h under dry argon to generate a poly(DMAPS) brush from the substrate (Figure 1), and was quenched by opening the vessel to air. The substrates were washed with hot water using a Soxhlet apparatus for 12 h.

2.3. Measurements. The thickness of the polymer brushes on the silicon substrate in air (the relative humidity = 19%) were determined by an imaging ellipsometer (Nippon Laser & Electronics Lab.) equipped with a YAG laser (λ = 532.8 nm). The polarizer angle was fixed at 50°, and a refractive index of 1.50 was used for calculations of the film thickness. Topographic imaging and force measurement were performed by AFM (SII Nanotechnology Inc., E-Sweep). The curvature radius of the used cantilever (Sphere Tips CONT, Nanosensors) made of SiO_2 was 1000 nm. The bending spring constant of the cantilever tip was 0.24 N/m. The cantilever surface was covered with *n*-propyltriethoxysilane (Sigma Aldrich Co.) monolayer by CVA method. A liquid cell was used for the measurement in pure water and 0.5 M NaCl aqueous solution at 298 K. NR measurements were carried out by a multilayer interferometer for neutrons (MINE) in JRR-3 at TOKAI, using wavelength of λ = 0.88 nm. The scattering vector, q, in specular reflectivity is defined by $q = (4\pi/\lambda)\ sin\theta$. The fitting curves of NR profiles were calculated by using Parratt32 software.

3. Results and Discussion.

The swollen thicknesses of poly(DMAPS) brush ($M_w \sim 3.2 \times 10^5$, $M_w/M_n \sim 1.8$) in aqueous solution were evaluated by AFM [7]. The interaction force of poly(DMAPS) brush compressed by a cantilever with radius R of probe head was measured in a pure water and in 0.5 M NaCl aqueous solution. The measured force F was converted to the force F/R. As shown in Figure 2, the F/R value increased with compressing the brush layer and with decreasing the distance from the substrate surface D. However, absolute distance from the substrate surface cannot be given directly by the force curve measurement. Let us define the gap D' as a distance between the position at the beginning of repulsive force detection and the constant compliance region in the force curve where the cantilever deflection was proportional to the piezo displacement. The offset distance D_0 was determined by scanning the scratched sample across the boundary of a scratched and an unscratched region. In this study, D_0 was obtained as a height of the brush layer from the scratched surface by AFM image taken under a constant force mode by applying the force corresponding to the constant compliance region in the force curve measurement. The equilibrium thickness L_e of the swollen polymer brush was defined by $L_e = D_0 + D'$. The thickness of poly(DMAPS) brush in air (relative humidity = 19%) was 147 nm, which was agreed with the value estimated by ellipsometry. The L_e values in pure water and in 0.5 M NaCl aqueous solution were 212 nm and 352 nm, respectively. We supposed that poly(DMAPS) formed relatively shrunk structure in pure water due to the attractive electrostatic interactions based on ammonium cations and sulfonyl anions, while these electrostatic interactions were screened by hydrated salt ions in 0.5 M NaCl aqueous solution. Poly(DMAPS) brush immersed in a salt solution behaves like an electrically neutral polymer to give relatively stretched chain structure due to the osmotic pressure. In the strict sense, the L_e estimated in this method cannot describe the true thickness of swollen brush, because the electrostatic interaction between the probe head and the outermost polyelectrolyte brush should be considered, especially in the case of a low ionic strength solution. Then, we tried to estimate the swollen thickness of poly(DMAPS) brush by NR, too.

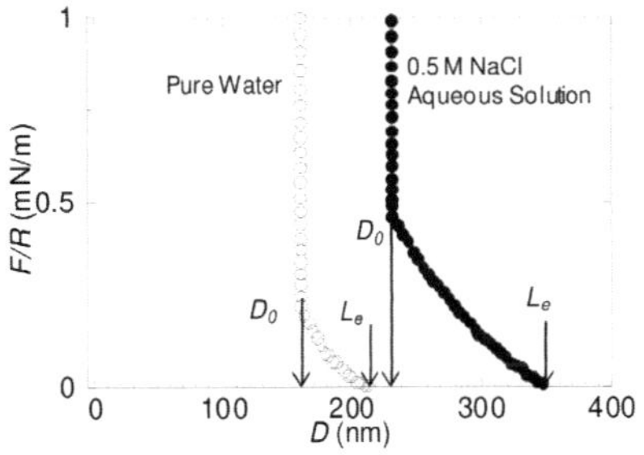

Figure 2. F/R vs D curves of the poly(DMAPS) brushes in pure water and 0.5 M NaCl aqueous solution. D_0 is evaluated by the height of scratched poly(DMAPS) brush to measure under a constant force corresponding to the constant compliance region. L_e is the critical distance at which a repulsive force is detectable.

Figure 3 shows the NR curves and scattering length density (SLD) profiles of poly(DMAPS) brush in D_2O and 0.5 M NaCl/D_2O. The neutron SLD of poly(DMAPS) brush in D_2O was dramatically increased from 4.75×10^{-4} to 6.32×10^{-4} nm^{-2} along with the distance from the substrate (z), and was slightly increased up to 6.39×10^{-4} nm^{-2} at the pure D_2O region. Theoretically, the SLD profiles are described by a parabolic function and an additional exponential tail at larger z region [13]. Figure 4 represents volume fraction (φ) curve of DMAPS segments as a function of z and the corresponding parabolic-type profile $\varphi = \varphi_0\{(1-z/h^*)^2\}$ assuming a brush layer height (h^*). In this study, we regard h^* as thickness of the swollen brush. The tail part in a volume fraction curve at $z > h^*$ region was caused by a roughness at the interface of brush and solution. The h^* for poly(DMAPS) brush in D_2O solution was 130 nm in pure D_2O. This result indicated that the brush chains were shrunk due to the attractive electrostatic interaction. On the other hand, the SLD profile in 0.5 M NaCl/D_2O drastically increased to 5.50×10^{-4} in 50 nm region from the substrate, and gradually increase to 6.29×10^{-4} nm^{-2} at the external solution interface. The h^* for poly(DMAPS) brush in 0.5 M NaCl/D_2O was estimated to be 280 nm. The moderate increment from 5.50×10^{-4} nm^{-2} to 6.29×10^{-4} nm^{-2} in SLD profile of 0.5 M NaCl/D_2O indicated that the brush chain was further extended from the quartz substrate because hydrated salt ions screened the attractive electrostatic interaction. The steep increment 4.30×10^{-4} nm^{-2} to 5.50×10^{-4} nm^{-2} was observed at z = 0 ~ 40 nm. We supposed that the brush chains located near the quartz substrate was not completely extended. Dimension of poly(DMAPS) brush in 0.5 M NaCl solution has been studied by Kato et al. using a light scattering. They reported that R_H of unbound poly(DMAPS) chain was continuously enlarged along with NaCl salt concentration from 0.06 M to 1.0 M [1]. Similar results were also obtained in the case of poly(DMAPS) chains adsorbed on a silica surface [4]. Therefore, 0.5 M of NaCl aqueous solution was not high concentration enough for free poly(DMAPS) to screen the intra or inter molecular electrostatic interaction completely. In addition, large molecular weight distribution of grafting polymer gives the gradient distribution of polymer concentration in the swollen brush layer with lower polymer density at the outermost surface and higher at the bottom region. The screening of electrostatic interaction would effectively take place at the external solution/brush interface forming relatively extended chain conformation, while the brush chains near the substrate surface remained shrinking.

Polymer chains at the outermost region were stretched up to $h^* = 280$ nm from the substrate. The thickness of brush layer was clearly increased with NaCl concentration in D_2O. We have to point out here that the h^* values determined by NR measurement were not agreed with the L_e values estimated by AFM. One of the reasons is supposed to be attributed to the molecular weight distribution. Considering the polymer chains with various molecular weights and/or length tethered on a flat substrate, the local concentration of polymer in the outermost surface region of the swollen brush must be much lower than that in the deeper region. The probe for the force curve measurement is so sensitive that the repulsive force can be detectable from the longest-length chain in the brush in spite of the low concentration at the outermost brush/solution interface region. Meanwhile, it is hard for NR measurement to determine the neutron SLD curve of the low polymer-concentration layer at the outermost brush/deuterated solvent interface because of the low H/D contrast. Therefore, we suppose that the larger thickness L_e was obtained from AFM compared with h^* estimated by NR measurement.

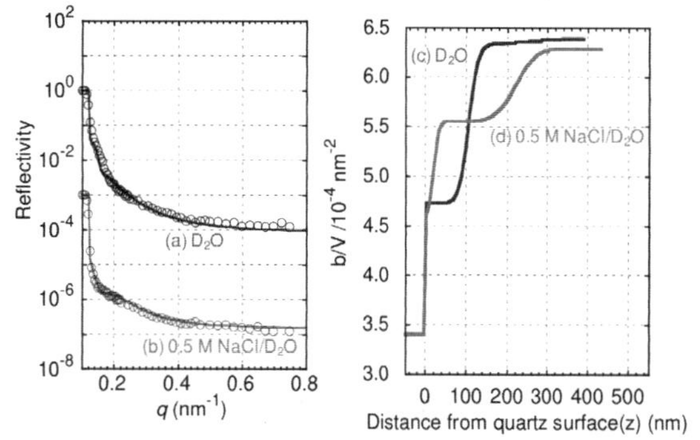

Figure 3. (a) NR curves of poly(DMAPS) brush/D_2O, (b) poly(DMAPS) brush/ 0.5 M NaCl in D_2O, and (c)-(d) their corresponding neutron scattering length density profiles along with the distance from quartz surface, respectively. Scattering vector $q = 4\pi \sin \theta/\lambda$ at $\lambda = 0.88$ nm.

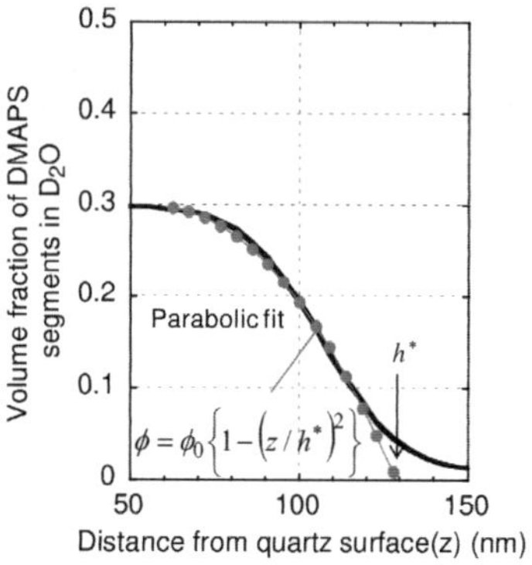

Figure 4. (a: open circle) Volume fraction distribution (φ) for poly(DMAPS) brush in D_2O calculated by Figure 3(c), and (b: filled circle) the corresponding fit based on a Parabolic function using $\varphi = \varphi_0\{(1\text{-}z/h^*)^2\}$.

4. Conclusion.

This is the first report on the analysis of the swollen poly(DMAPS) brush structure in aqueous solution with different salt concentrations by AFM and NR measurement. The equilibrium thickness L_e of the swollen polymer brush in pure water evaluated by force curve measurement was increased from 212 to 352 nm by addition of salt to the solution. The hydrated salt ions screened the strong attractive electrostatic interaction of sulfobetaine groups to give the structure change of the brush from shrunk to relatively stretched state. Similar swelling state was observed by NR measurement, too. The neutron SLD profile represented the gradient structure of swollen brush due to the large molecular weight distribution.

Acknowledgments

The authors gratefully acknowledge the financial supports by a Grant-in-Aid for the Global COE Program, "Science for Future Molecular Systems" from MEXT, Japan.

References

[1] Kato T and Takahashi A 1996 *Ber. Bunsenges. Phys. Chem.* **100** 784.
[2] Liaw D -J, Lee W -F, Whung Y -C and Lin M -C 1987 *J. Appl. Polym. Sci.* **34** 999.
[3] Schulz D N, Peiffer D G, Agarwal P K, Larabee J, Kaladas J J, Soni L, Handwerker B and Garner R T 1986 *Polymer* **27** 1734.
[4] Kato T, Kawaguchi M and Takahashi A 1999 *Langmuir* **15** 4302.
[5] Zhang Z, Chao T, Chen S and Jiang S 2006 *Langmuir* **22** 10072.
[6] Cheng N, Brown A A, Azzaroni O and Huck W T S 2008 *Macromolecules* **41** 6317.
[7] Yamamoto S, Ejaz M, Tsujii Y, Matsumoto M and Fukuda T 2000 *Macromolecules* **33** 5602.
[8] Kobayashi M, Terayama Y, Hosaka N, Kaido M, Suzuki A, Yamada N, Torikai N, Ishihara K, and Takahara A 2007 *Soft Matter* **3** 740.
[9] Ohno K, Morinaga T, Koh K, Tsujii Y and Fukuda T 2005 *Macromolecules* **38** 2137.
[10] Kobayashi M and Takahara A 2005 *Chem. Lett.* **34** 1582.
[11] Duann Y F, Chen Y C, Shen J T and Lin Y H 2004 *Polymer* **45** 6839.
[12] Hayashi K, Saito N, Sugimura H, Takai O and Nagagiri N 2002 *Langmuir* **18** 7469.
[13] Milner S T, Witten T A and Cate M E 1988 *Macromolecules* **21** 2610

Journal of Physics: Conference Series **184** (2009) 012012 doi:10.1088/1742-6596/184/1/012012

Combinatory Usage of X-ray and Neutron Diffraction Techniques for the Refined Structure Analysis of Polymer Crystals: from Hydrogen Atoms to Bonded Electron Density Distribution

Kohji Tashiro[1], Makoto Hanesaka[1], Yoshinori Yoshizawa[1], Takashi Ohhara[2] Ryota Kuroki[2], Toshiaki Kitano[3], Takashi Nishu[3], Sono Sasaki[4] and Masaki Takata[4]

[1]Toyota Technological Institute, 2-12-1, Hisakata, Tempaku, Nagoya, 468-8511, Japan.

[2]Japan Atomic Energy Agency, Shirakta-Shirane, 2-4, Tokai, Ibaraki, 319-1195, Japan.

[3]Nagoya University, Frou, Chikusa, Nagoya, 464-8603, Japan.

[4]JASRI, Spring-8, 1-1-1, Kouto, Mikazuki, Sayo, Hyogo, 679-5198, Japan.

[1]E-mail: ktashiro@toyota-ti.ac.jp.

Abstract. In order to predict the physical property of polymers, we need to know the atomic coordinates including hydrogen atoms as accurately as possible. However, polymer samples give us limited number of broad X-ray reflections in general. In order to improve this problem, we have made many efforts to collect the X-ray diffraction data as accurate as possible. In this paper, we report the structure analysis including the extraction of hydrogen/deuterium atomic positions for the oriented polyoxymethylene by the combination of X-ray diffraction method with neutron diffraction method. In such a cases of polydiacetylene giant single crystal the bonded electron density distribution was evaluated along the skeletal chain by the so-called X-N method based on the wide-angle X-ray diffraction data and the wide-angle neutron diffraction data.

1. Introduction

In order to predict the ultimate physical properties of polymer crystals, we need to know the molecular interactions and atomic coordinates including hydrogen atoms which play an important role in giving the anisotropy in the physical property [1]. However, general polymers give us only small number of broad X-ray reflections, making it difficult to analyze the crystal structure with high reliability. We have made many efforts to increase the reliability as listed below.

(1) Utilization of an X-ray beam with short wavelength to increase the total number of observable reflections [2].

(2) Development of a software to separate the overlapped reflections and evaluate the integrated intensity of separated reflections accurately [2].

(3) Utilization of energy calculation technique to extract energetically acceptable structure models, which can be used as the initial model for the structure analysis [3].
(4) Extraction and refinements of hydrogen atomic positions utilizing electron diffraction and wide-angle neutron diffraction (WAND) methods [4].
(5) Utilization of giant single crystals of polymers [5].

In an ideal case many reflections can be collected from a giant single crystal of polymer, and it becomes possible to clarify the bonded electron density distribution on the basis of reasonable combination of X-ray and neutron data (X-N method). In the present paper, we report the results of structure refinement made for polyoxymethylene (POM) and polydiacetylene on the basis of diffraction peaks obtained as many as possible.

2. Experimental Section

Normal POM samples used in the experiment were prepared by γ-ray induced solid-state polymerization reaction of tetraoxane single crystals. The deuterated trioxane crystal gave the deuterated POM, $-(CD_2\text{-}O)_n-$, after the γ-ray polymerization reaction. The giant single crystals of polydiacetylene with bulky side groups, poly [1.6-di(*N*-carbazoyl)-2,4-hexadiyne] (PDCHD), were synthesized from a single crystals of the corresponding diacetylene monomer by irradiating γ-ray beam.

The X-ray diffraction data were collected using Rigaku Rapid-II diffractometer with Mo-K_α beam. In parallel, the high-energy synchrotron X-ray diffraction measurement was performed in the beam line 04B2 of SPring-8, Japan. The WAND measurements of hydrogenated and deuterated POM samples and normal PDCHD crystal were performed using an imaging plate system BIX-3 [6] installed in JRR-3 of Japan Atomic Energy Agency (Tokai-mura).

3. Results and Discussion

3.1. Structure Refinement of POM

The crystal structure of POM had been reported by many researchers [7 - 13]. Several chain models containing some conformational distortions were proposed. But, more detailed structure including the hydrogen atomic positions was not yet reported satisfactorily. The X-ray fiber diagram of POM taken

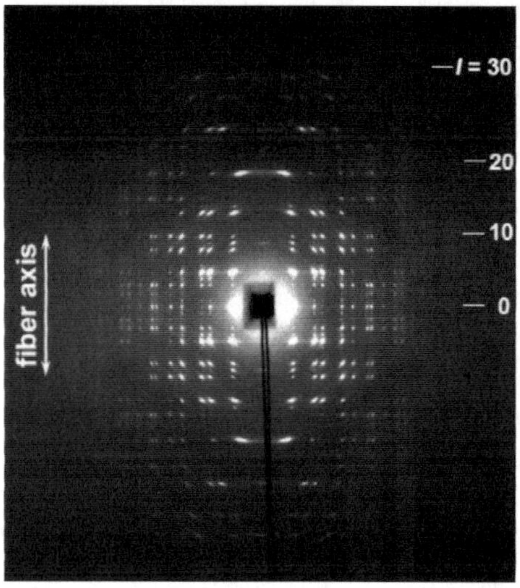

Figure 1. WAXD data of POM sample measured with synchrotron X-ray beam at BL04B2 of SPring-8.

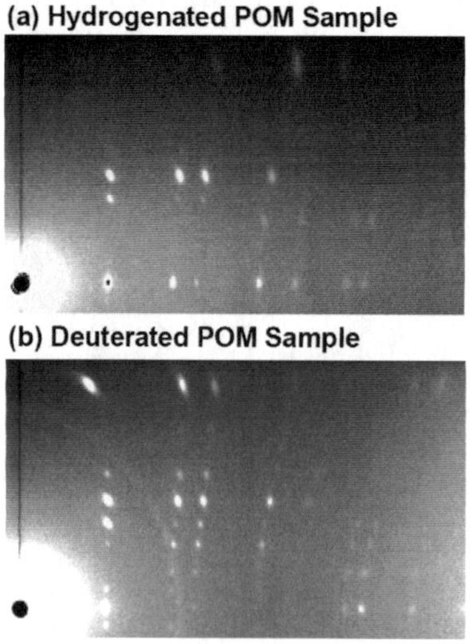

Figure 2. WAND data of (a) hydrogenated and (b) deuterated POM samples measured at room temperature. The vertical direction is parallel to the fiber axis.

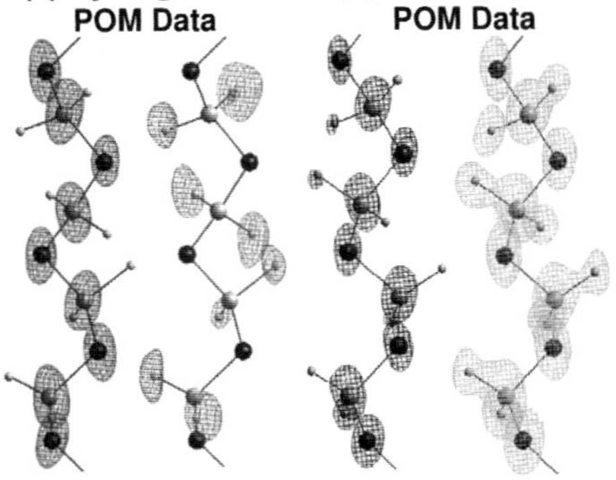

Figure 3. 3-dimensional Fourier maps (F_o) of (a) hydrogenated POM sample and (b) deuterated POM sample obtained from the WAND data at room temperature. $F_o(+)$ and $F_o(-)$ indicate the density distribution of atomic nuclei with the positive and negative density height, respectively. $F_o(+$ high) and $F_o(+$ low) are the contour maps viewed at the different heights from the baseline.

at -150 °C using a synchrotron X-ray source of 0.328 Å wavelength is shown in Figure 1. From this pattern, 700 reflections were collected, in which the unique reflections were 202. The unit cell parameters were $a = b = 4.464$ (4) Å, c (*fiber axis*) = 17.389 (2) Å and $\gamma = 120^o$ at room temperature, and $a = b = 4.373$ (4) Å, c (*fiber axis*) = 17.274 (2) Å and $\gamma = 120^o$ at -150 °C. The space group is $P3_2$. Based on thus-collected large number of reflections, the initial structure model and the hydrogen atomic positions were extracted successfully by a direct method and also by the differential Fourier synthesis. The refinement was made for all the coordinates of carbon, oxygen and hydrogen atoms by taking the anisotropic (carbon and oxygen atoms) and isotropic thermal parameters (hydrogen atoms) into consideration. The final reliability factors were 6.9 % at -150°C and 8.8 % at room temperature. The 2-dimensional WAND patterns taken for hydrogenated and deuterated POM samples are shown in Figure 2. The total number of unique neutron reflections were 28 and 72 reflections, respectively. The structure refined by the X-ray analysis was transferred to the WAND analysis. The final reliability factors were 13.9 % for hydrogenated POM and 22.4 % for deuterated POM. The hydrogen atomic positions were extracted successfully as the negative and positive peaks in the 3-dimensional Fourier maps as shown in Figure 3. More accurate analysis of synchrotron X-ray diffraction data of POM gave us more ore less different structure model from the above-mentioned (9/5) helical model, in which the (29/16) helices are packed in the cell. The reliability factor calculated for this model (8.6 %) is almost the same as that calculated for the (9/5) helices. But, the existence of a series of *00l* reflections, which can not be predicted for the (9/5) helical model, was confirmed on the Weissenberg photograph and supports the (29/16) helical structure with 55.7 Å period.

3.2. Bonded Electron Density Distribution of PDCHD

Figure 4 shows the molecular chain conformation analyzed using WAXD data. The main chain of planer-zigzag form consists of an alternation of single, double and triple bonds. An X-ray beam is scattered by the electron clouds around the atomic nuclei and the WAXD analysis gives us the information on electron density distribution. The neutron beam is scattered by the nuclei, and the nuclear density distribution information is obtained. By utilizing both of these two kinds of diffraction data, we may find the bonded electron density distribution (X-N method) [14]. Figure 5 shows the thus-obtained bonded electron density distribution map along the main chain of PDCHD. The electron density around the triple bonds shows the highest density compared with those around the single and double bonds. But, the electron densities of single bonds and double bonds are almost the same height as known from Figure 5, suggesting a development of some degree of electronically conjugated system along the skeletal chain. In order to confirm the result, the electron density distribution of PDCHD chain was calculated on the basis of a density functional theory (DMol3, Material Studio Version 4.2 (accelrys)). The calculated bonded electron density distribution is consistent qualitatively at least with the experimental result.

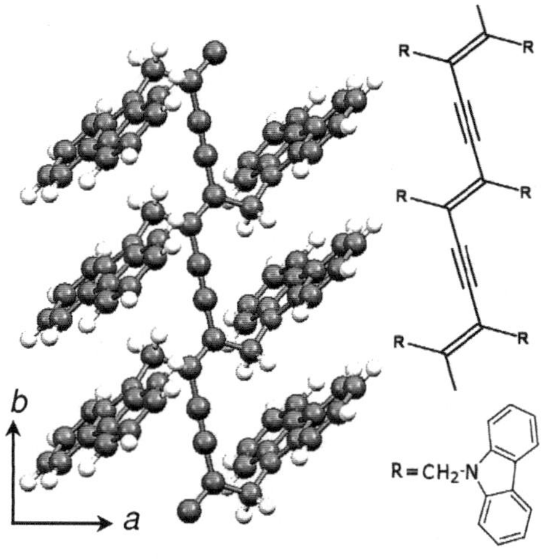

Figure 4. Molecular chain conformation of PDCHD.

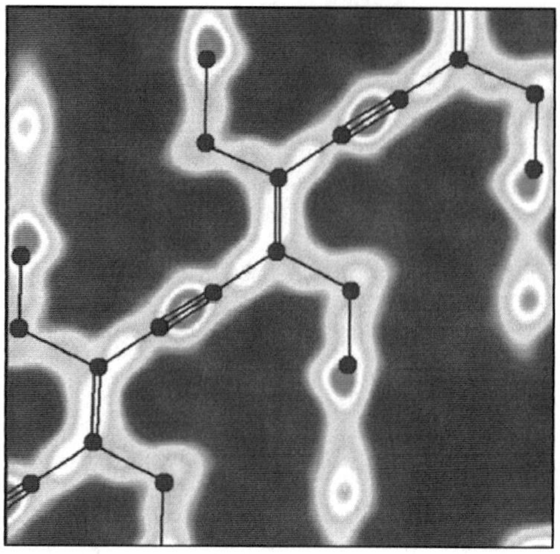

Figure 5. Bonded electron density distribution along PDCHD chain by the X-N method based on WAXD and WAND data.

4. Conclusions

In the present paper, we have reviewed the structure analysis results of two types of polymer crystals, POM and PDCHD on the basis of organized combination of X-ray and neutron diffraction techniques. The increased total number of observed reflections has allowed us to obtain the hydrogen atomic positions for POM and the bonded electron density distribution for PDCHD. But, these two examples are lucky cases with great amount of observed reflections. The general synthetic polymers give only limited number of reflections and are still at the stage in difficulty of the refined structure analysis. We need to apply the ideas listed in the introductory section to the general cases.

References

[1] Tashiro K 1993 *Prog. Polym. Sci.* **18** 377-435
[2] Tashiro K Asanaga H Ishino K Tazaki R and Kobayashi M 1997 *J. Polym. Sci. Part B Polym. Phys. Ed.* **35** 1677-1700
[3] Tashiro K Yoshino J Kitagawa T Murase H and Yabuki K 1998 *Macromolecules* **31** 5430-5440
[4] Tashiro K Tanaka I Ohhara T Niimura N Fujiwara S and Kamae T 2004 *Macromolecules* **37** 4109-4117
[5] Tashiro K Nishimura H and Kobayashi M 1996 *Macromolecules* **29** 8188-8196
[6] Niimura N Karasawa Y Tanaka T Miyahara J Akahashi K Saito H Koizumi S Hidaka M 1994 *Nucl. Instrum. Methods* **A349** 521-525
[7] Tashiro K Hanesaka M Ohhara T Ozeki T Kitano T Nishu T Kurihara K Tamada T Kuroki R Fujiwara S Tanaka I and Niimura N 2007 *Polymer J.* **39** 1253-1273
[8] Staudinger H Johner H Signer R Mie G and Hengstenberg J 1927 *Z. Physik. Chem.* **126** 425
[9] Tadokoro H Yasumoto T Murahashi S and Nitta I 1960 *J. Polym. Sci.* **44** 266-269
[10] Uchida T and Tadokoro H 1967 *J. Polym. Sci. Part A-2* **5** 63-81
[11] Carazzolo G A 1963 *J. Polym. Sci. Part A* **1** 1573-1583
[12] Takahashi Y and Tadokoro H 1978 *J. Polym. Sci. Polym. Phys. Ed.* **16** 1219-1225
[13] Saruyama Y Miyaji H and Asai K 1979 *J. Polym. Sci. Polym. Phys. Ed* **17** 1163-1170
[14] Wilson C C 2000 *Single Crystal Neutron Diffraction from Molecular Materials* (World Sci. Pub. Co. Pte. Ltd., Singapore)

Nanoporous structures templated from block copolymer morphologies

Hideaki Yokoyama,[*,†] Lei Li,[‡] Cedric Dutriez,[‡] and Kenji Sugiyama[§]

[†]Graduate School of Frontier Sciences, The University of Tokyo, 5-1-5 Kashiwa-no-ha, Kashiwa, Chiba 277-8561, Japan, [‡]Nanotechnology Research Institute, National Institute of Advanced Industrial Science and Technology, Central 5, Higashi, 1-1-1, Tsukuba, Ibaraki 305-8565, Japan, [§]Department of Organic and Polymeric Materials, Tokyo Institute of Technology, 2-12-1, Ohokayama, Meguro-ku, Tokyo 152-8552, Japan,

*email: yokoyama@molle.k.u-tokyo.ac.jp

Abstract. We studied nanoporous structures templated from block copolymer morphologies using carbon dioxide swelling and de-swelling. The volume where carbon dioxide occupied turns into empty space after de-swelling. As a pressure of carbon dioxide increases, the effective volume fraction of block copolymer changes due to the selectivity of carbon dioxide. The swollen morphology of block copolymer changes from its original neat morphology and a variety of nanoporous structures appear after de-swelling. Consequently, unique nanoporous stuructures depending on the process pressure were obtained.

1. Introduction

Block copolymers have been studied primarily due to their unique self-assembled morphologies with a size scale of tens of nanometers. For diblock copolymers, which have the simplest architecture among various kinds of block copolymers, spheres with body centered cubic symmetry, cylinders with hexagonal symmetry and one dimensionally periodic lamellae have been known as thermodynamically allowed morphologies for decades. In the late 1990's, a bicontinuous cubic phase known as double gyroid was added to the equilibrium morphologies.

Fabrication of nanostructures has been receiving strong attention in fundamental science as well as in industrial applications.[1,2] However, large-area structured surfaces and thin films on the nanometer scale still remains a challenge for conventional fabrication methods. In order to overcome the difficulties of large-area production, molecular self-assembly has been considered to be a serious candidate. One of such self-assembled materials is block copolymer, which has been considered for nano-technological applications not only due to the scale of the nanodomains, but also the convenient tunability of size, shape, and periodicity by changing their molecular structures.[3-6] However, available size, shape and symmetry in block copolymer templating are restricted by the molecular structures; therefore, fine-tuning requires precisely controlled polymerization.

Supercritical carbon dioxide is an inexpensive, non-toxic and non-flammable fluid, which was widely applied for materials synthesis and process.[7] Gas-like diffusivities, the continuously tunable solvent power and selectivity, and the complete elimination at the end of the process make carbon dioxide very attractive.[8] Particularly, carbon dioxide has been used to make microcells, cells with a diameter of the order of micrometer.[9-11] However, reducing the size and obtaining non-spherical shapes have been a big challenge. Our group has, nevertheless, demonstrated that carbon dioxide can be used to fabricate nanometer scale cells using fluorinated block copolymers as templates for CO_2 foams. CO_2 was used as a porogen to introduce nanometer cells and pores in the nanodomains of bulk and thin film of block copolymers having fluorinated blocks.[12-15] The key to the successful introduction of nanometer cells and pores is selective swelling of the CO_2-philic fluorinated block domains with CO_2. After complete removal of CO_2, the volume occupied with CO_2 turns into empty voids. The size of cells can be tuned by changing process conditions with the same template. Moreover, this methodology can be applied to thin films [13,14] as well as monoliths and also to another sphere forming non-fluorine bcp, a block copolymer of polystyrene and CO_2-philic polydimethylsiloxane.[16] Here we report that selective swelling with CO_2 of bcp thin films induces morphological changes and results in unique non-spherical nanostructures that differ from the original bcp morphology.

2. Experimental

2.1. Polymer Synthesis and Sample Preparation.

We synthesized polystyrene-*b*-poly(perfluorooctylethyl methacrylate) (PS-PFMA) with two different block fractions by sequential living anionic polymerization. The molecular weights of blocks in two PS-PFMAs are 20,000-13,000 and 10,000-7,670 g mol^{-1}.[15] The molecular weights of the PS blocks were determined by gel permeation chromatography, and the molecular weights of the PFMA blocks were determined by ^{1}H-nuclear magnetic resonance. Two PS-PFMAs were mixed in order to adjust the average volume fraction of PFMA to 33 vol% and named as PSPFMA33.

α,α,α-trifluorotoluene (Aldrich) was used as spin-casting solvents, and were named as TFT. 500-nm-thick PS-PFMA films were prepared by spin-casting of those solutions onto silicon wafers. The silicon wafers (Shinetsu Co.) were used as received.

2.2. scCO_2 Process.

A stainless steel high-pressure vessel for CO_2 processes was connected to a high-pressure liquid chromatography pump (JASCO PU-2086 plus) with a cooling head and to a back-pressure regulator (JASCO SCF-Bpg). Films of PS-PFMA were placed in the high-pressure vessel at 60 °C for 30 min with a constant pressure. The vessel was placed in an ice bath to quench the films to 0 °C while maintaining the pressure using the pump and regulator. After the temperature and pressure were stabilized, pressure was released. The depressurization rate was controlled at 0.5 MPa/min.

2.3. Spectroscopic Ellipsometry.

Film thickness and refractive index were measured with a JASCO M-220 ellipsometer with incident light in a wavelength range of 400-800 nm at an incident angle of 60° with respect to the surface normal. The refractive indices of the nanocellular thin films were computed assuming constant refractive indices throughout the films.

2.4. Reactive Ion Etching.

The etching process to expose embedded nanostructures to the surface was performed with a reactive ion etcher (SAMCO compact etcher FA-1) with a CF_4 flow rate of 2 mL/min, a pressure of 10 Pa, and a power density of 10 W/cm^2. Under this condition, the etching rate for PS-PFMA copolymer thin films is ~1 nm/s.

2.5. Scanning Electron Microscopy.

Films were fractured in the liquid nitrogen in order to expose the internal structures to the surface. The cyro-fractured surface was obtained using an environmental scanning electron microscope (ESEM) (Philips XL20 ESEM-FEG) equipped with a field-emission gun under high-vacuum conditions. A 10 keV electron beam was used for the observation. No conductive coatings were applied onto the specimen surfaces.

2.6. Grazing incident small angle X-ray scattering (GISAXS).

GISAXS experiments were conducted at 15A beamline of Photon Factory in KEK. A sample was placed in a vacuum chamber with Kapton windows on a computer controlled gonio stage. The X-ray beam impinges at a small incident angle α onto a thin film supported on a substrate. The scattered X-ray was detected with an image intensifier accompanied with a two dimensional charge coupled device camera. Experimental detail of GISAXS at SPring-8 can be found elsewhere[13]. The wave vectors are defined as $q_y=(2\pi/\lambda)\sin\theta\cos\beta$ and $q_z=(2\pi/\lambda)(\sin\alpha+\sin\beta)$, where θ is the inplane angle, α and β are incident and scattered vertical angles to the surface plane, respectively, and λ is a wavelength (1.5Å).

3. Results and Discussion

The mixture of PS-PFMAs described in the Experimental section was dissolved in TFT and then spin-casted onto silicon wafers. A typical thickness of those films was 500 nm, which was measured by ellipsometry. The thin films were then placed in a high-pressure vessel and pressurized with carbon dioxide as described in the experimental section for 30 minutes at a various pressure and at 60 °C. Then the whole vessel was placed in an ice bath to reduce temperature down to 0 °C and to freeze the swollen morphology by the glass transition of PS in carbon dioxide. Carbon dioxide pressure was then slowly reduced to remove carbon dioxide without destroying the structures of the block copolymer.

Previously, we reported that this mixture of block copolymers presents the nanosheet structure parallel to the substrate when processed with 8 MPa of carbon dioxide.[15] We further increased the carbon dioxide pressure to swell the block copolymer film and then removed carbon dioxide. An example of the SEM images of the films processed at 10 MPa of carbon dioxide is shown in Figure 1. The thin film on substrate was scratched at liquid nitrogen temperature and processed with RIE to provide a cross-sectional view of the films. Similar to previously obtained nanosheet structure, we again find nanosheets stacking on silicon substrate. However, the morphology is not as homogeneous as that at 8 MPa [15] and many defects are observed. It should be noted that SEM does not observe the domains of block copolymers due to lack of electron density contrast between the domains but observe the topographical structures on the surface. Therefore, the structures here are planar pores or nanosheets stacking on a substrate. Such structure is confirmed with the GISAXS pattern shown in Figure 1b. All the Bragg peaks appear on the q_z axis, which indicate the periodic structure perpendicular to the surface. The peaks are assigned to two sets of integer multiples of the first order peaks of direct and reflected components. The GISAXS pattern of this film is essentially the same as in that in our previous publication [15], where detailed peak assignment can be described.

We further increased the carbon dioxide pressure and found a significantly different structure at 22 MPa. An example of SEM image of the top surface of such structure and its GISAXS pattern are shown in Figure 2a and b. A top layer of nanosheets stacking on a substrate is observed and the orientation of nanosheets is similar to the cross-sectional image of the same copolymer processed at 10 MPa in Figure 1. However, we clearly find that a lot of cylindrical rods are connected to the nanosheets. GISAXS pattern in Figure 2b also shows the disappearing Bragg peaks on the q_z axis, which clearly indicates layered sheets are broken down. The nanosheets are apparently transformed into cylindrical rods similar to the order-to-order transition (OOT) of block copolymer morphology from lamellae to hexagonal cylinders. Although we don't find any order in the cylindrical rods, the shape of the nano-sized objects obeys the trend of OOT in block copolymers. It is interesting to find that a plane of nanosheets is broken down into cylindrical rods. The structure apparently is in the course of transition and is not in equilibrium. This could be due to the fairly high molecular weight of block copolymers, which reduces the mobility and slows down the OOT process. There is another possibility of such a mixed structure caused by the mixed block copolymers to adjust the volume fraction. We further increased pressure and checked if this is in the course of OOT or phase separation of two different molecular weight block copolymers showing two different morphologies, i.e. lamellae and cylinders

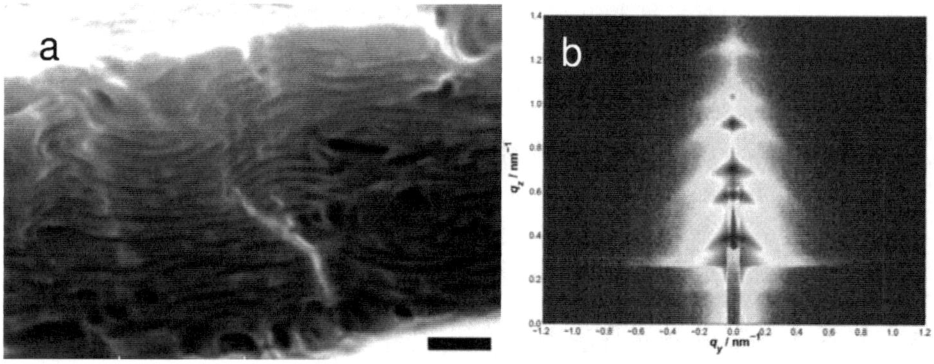

Figure 1. (a) Cross-sectional view of PSPFMA33 thin films casting from TFT after the $scCO_2$ process at a saturation pressure of 10 MPa and etching. The bar indicates 200 nm. (b) GISAXS pattern of PSPFMA33 thin films casting from TFT after the $scCO_2$ process at a saturation pressure of 10 MPa. An incident angle was set to 0.2 degree.

As a pressure of saturation increases to 30 MPa, we found that the structure completely changed to cylindrical rods and the nanosheets completely disappeared as shown in Figure 3a. GISAXS pattern in Figure 3b shows no Bragg peaks of the layered sheet structure. Only the strong form factor near the rectangular beam stopper is found. These results suggest that the morphology changed from lamellae to cylinders during the carbon dioxide process and, after carbon dioxide removed, cylindrical rods appeared. It is interesting to note that those cylinders are not ordered at all. They even show braches and form a network. We don't understand the reason at this point since the structural observation is limited after carbon dioxide is removed. Further in-situ study is necessary to fully understand those phenomena.

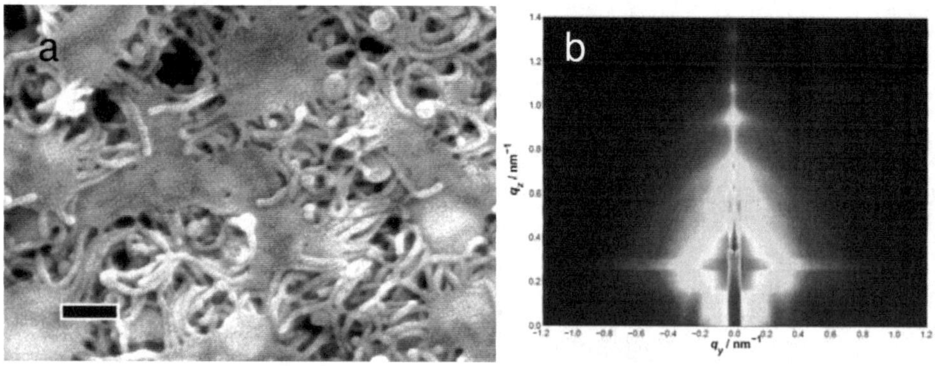

Figure 2. Top view of PSPFMA33 thin films casting from TFT after the $scCO_2$ process at a saturation pressure of 22 MPa. The bar indicates 200 nm. (b) GISAXS pattern of PSPFMA33 thin films casting from TFT after the $scCO_2$ process at a saturation pressure of 22 MPa. An incident angle was set to 0.2 degree.

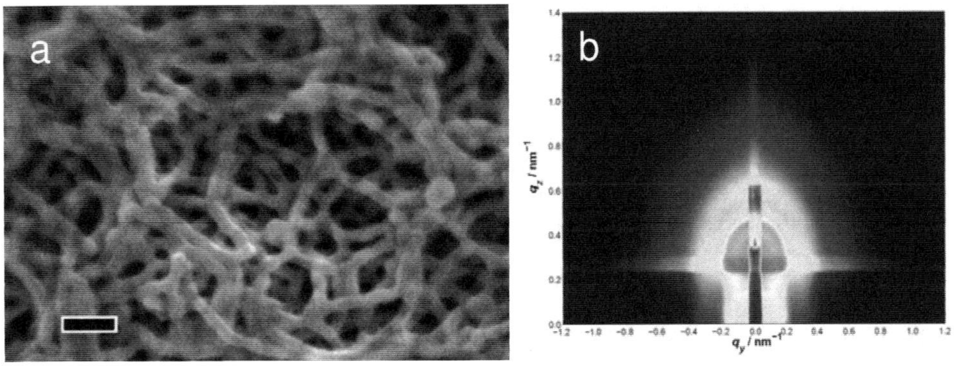

Figure 3. Top view of PSPFMA33 thin films casting from TFT after the $scCO_2$ process at a saturation pressure of 30 MPa. The bar indicates 100 nm. (b) GISAXS pattern of PSPFMA33 thin films casting from TFT after the $scCO_2$ process at a saturation pressure of 30 MPa. An incident angle was set to 0.2 degree.

Acknowledgment This work has been supported by Grant-in-Aid for Scientific Research on Priority Area "Soft Matter Physics" (No. 463) from the Ministry of Education, Culture, Sports, Science and Technology.

References

[1] Hamley I W 2003 *Angew. Chem. Int. Ed.* **42** 1692.
[2] Park C, Yoon J and Thomas E L 2003 *Polymer* **44** 6725.
[3] Fink Y, Winn J N, Fan S, Michel J, Joannopoulos J D and Thomas, E. L. 1998 *Science* **282** 1679.
[4] Thurn-Albrech T, Schotter J, Kastle G A, Emley N, Shibauchi T, Krusin-Elbaum L, Guarini K, Black C T, Tuominen M T and Russell T P 2000 *Science* **290** 2126.
[5] Pai R A, Humayun R, Schulberg M T, Sengupta A, Sun J N and Watkins J J 2004 *Science* **303** 507.
[6] Kim D H, Kim S H, Lavery K and Russell T P 2004 *Nano. Lett.* **4** 1841.
[7] DeSimone J M 2002 *Science* **297** 799.
[8] Reverchon E and Adami R 2006 *J. Supercrit. Fluid* **37**, 1.
[9] Zhang H F, Long J and Cooper A I 2005 *J. Am. Chem. Soc.* **127** 13482.
[10] Cooper A I 2003 *Adv. Mater.* **15** 1049.
[11] Cooper A I 2000 *J. Mater. Chem.* **10** 207.
[12] Yokoyama H., Li L, Nemoto T and Sugiyama K 2004 *Adv. Mater.* **16** 1542.
[13] Li L, Yokoyama H, Nemoto T and Sugiyama K 2004 *Adv. Mater.* **16** 1226.
[14] Li L, Nemoto T, Sugiyama K and Yokoyama H 2006 *Macromolecules* **39** 4746.
[15] Yokoyama H, Li L, Dutriez C, Iwakura Y, Sugiyama K, Masunaga H, Sasaki S and Okuda H 2008 *Macromolecules* **41** 8626.
[16] Li L and Yokoyama H 2006 *Angew. Chem. Int. Ed.* **45** 6338.

Journal of Physics: Conference Series **184** (2009) 012014 doi:10.1088/1742-6596/184/1/012014

Correlated Changes in Meso- and Nano-Scale Hierarchical Structure of Vinylidene Fluoride-Trifluoroethylene Random Copolymers Investigated by Simultaneous Measurements of DSC and SAXS/WAXD Utilizing Synchrotron Radiation

H Masunaga[1,*], S Sasaki[1], M Takata[1,2,3], M Hanesaka[4], K Tashiro[4]

[1]Japan Synchrotron Radiation Research Institute (JASRI) / SPring-8, Sayo, Hyogo 679-5198, Japan;
[2]The RIKEN Harima Institute / SPring-8, Sayo, Hyogo 679-5198, Japan;
[3]Graduate School of Frontier Science, The University of Tokyo, Chiba 277- 8561, Japan;
[4]Department of Future Industry-oriented Basic Science and Materials, Graduate School of Engineering Toyota Technological Institute, Tempaku Nagoya 468-8511, Japan.
E-mail: masunaga@spring8.or.jp

ABSTRACT. Hierarchical structural features of Vinylidene fluoride-trifluoroethylene (VDF-TrFE) copolymers during the phase transition have been successfully investigated in detail by *in-situ* simultaneous measurements of DSC and SAXS/WAXD utilizing synchrotron radiation. In a heating process at a constant rate of 1 °C /min, a broad endothermic peak was measured for a VDF-TrFE copolymer from ca. 75°C to ca. 135°C. On the other hand, WAXD data indicated that the phase transition from the crystal form II to the high-temperature (HT) phase began to occur at ca. 105°C. With an increase in temperature from ca. 75° C to ca. 105° C (temperature region A), lamellar staking distance (long period) and lamellar thickness evaluated in SAXS data analyses slightly increased though the crystallite size (the lateral dimension of the lamellae) was almost constant. The (110) reflections of the crystal form II gradually decreased in integrated intensity and scattering from the amorphous region increased instead, which implied that lamellae in crystal form II became partially disordered before the phase transition.

1. Introduction

A part of vinylidene fluoride-trifluoroethylene (VDF-TrFE) random copolymers exhibits the phase transition between the crystal form II and the high-temperature (HT) phase at Curie transition temperature.[1-3] The crystal form II consists of the packing in which chain conformation is the trans-gauche(+)-trans-gauche(-). The HT phase consists of the molecular chains with statistical distribution of trans and gauche bonds along the chain axis, and these chains rotate violently around the chain axis due to the thermally activated motions. In the phase transition temperature region where the trans-to-gauche conformational change of chains occurs in the crystalline lattice, the thickness of stacked lamellae drastically increased. However, the relationship between the nano- and meso-scale structural changes has not been revealed in detail.

The purpose of this study is to clarify the hierarchal structure change connected with so-called crystallographic phase transition of a VDF-TrFE random copolymer. Differential scanning calorimetrory (DSC) can confirm temperature ranges of phase transitions. By small-angle X-ray scattering (SAXS) and wide-angle X-ray diffraction (WAXD) measurements utilizing synchrotron radiation, lamellar and crystal structures can be investigated in the *in-situ* heating process, respectively. Therefore, DSC/SAXS/WAXD simultaneous measurements were efficient for clarifying the entire structural change from the nano scale to the meso scale in the so-called phase transition correlation between a hierarchal structure change and a heat flow. It is necessary to utilize synchrotron radiation for tracing the details of structural changes. In the present paper we will report the experimental results on the melt-quenched sample of VDF 82% copolymer. We paid attention to slight changes on each hierarchal structure scale in the beginning of the first endothermic peak in a heating process, and considered the phenomenon to relate with chain activation.

2. Experimental Section

2.1. Sample:

Vinylidene fluoride–trifluoroethylene random copolymer of VDF 82 molar % used as a sample. The sample was supplied by Daikin Kogyo Co., Japan. The sample was melted at 200°C and rapidly quenched into liquid nitrogen, which contained mostly the crystal form II. The thus melt-quenched VDF 82% was used for measurements.

2.2. DSC/SAXS/WAXD Measurements:

DSC, SAXD and WAXS measurements were performed simultaneously at BL40B2 beamline in SPring-8, Japan[4]. The wavelength of incident X-ray beam used in this study was 1.5 Å. The 2D-WAXD pattern was taken with a BAS-MS2025 imaging plate (IP, FUJIFILM Corp., Japan) set into an original IP holder. The 2D-SAXS signal from the sample passed through a hole with an 8 mm diameter, which was opened at the center of IP for WAXD, and it was detected with an IP installed in an R-AXIS IV++ IP System (Rigaku Co., Japan). An FP84HT hot stage where the detector-side window of the hot stage was customized in opening an angle for measuring the X-ray scattering up to the scattering angle 28° and an FP90 control unit (Mettler-Toledo International Inc., U.S.A.) were utilized for DSC measurements. The DSC/WAXD/SAXS simultaneous measurement was performed for every 5 min at a temperature increased by 5°C step in the heating process. The DSC thermogram was started to be recorded at a heating rate of 1°C/min from ca. 30°C.

The SAXS data were analyzed on the basis of the two-phase model consisting of crystal and amorphous phases using the one-dimensional electron density correlation function of the SAXS data.[5] A long period which is a lamination interval of lamellar crystals and a crystal thickness estimated here were defined L and d_c, respectively. The crystallite size was estimated from the reflection width on the basis of Scherrer's equation, where the instrumental line width was estimated on the basis of the diffraction data of CeO_2 powder (NIST standard sample).

3. Results and Discussion

The change in the lamellar morphology was monitored by the SAXS measurements. The temperature dependence of the SAXS profiles corrected for the Lorentz factor ($I(q)q^2$) at 40.5, 105.5, 125.5, 145.5, 155.5°C is shown in Figure 1. Here, q is the magnitude of the scattering vector, and q is $4\pi\sin\theta/\lambda$, where 2θ is a scattering angle, and λ is a wavelength. The peak observed at about q=0.2 nm^{-1} corresponds to a long period. The peak position was sift slightly to a lower angle and the intensity increased under 145.5°C. The peak position largely shifted to a low angle upper 145.5°C, and the peak disappeared at 155.5°C. It is shown that a long period spread gradually under 145.5°C and spread rapidly near 155.5°C.

Figure 2 shows that the WAXD profiles measured for the sample in the heating process. The peaks about 2θ=16, 17, 18, 25, 25.5° are assigned to the reflections of $(010)_{II}$, $(200)_{II}$, $(110)_{II}$, $(201)_{II}$,

$(111)_{II}$ by assuming the orthodontic unit cell for a crystal form II. [6-8] As the temperature is increased, the intensity for the reflections of $(110)_{II}$ and $(201)_{II}$ decreased remarkably at about 125.5°C. The peak near $2\theta = 17^{o}$ shifted to lower angle side with slight change in intensity. In higher temperature range only this peak was observed and disappeared above the melting point. As the temperature is increased, the integral peak intensity near $2\theta = 18^{o}$ became small. This indicates that the crystal form II transforms to the HT phase near 125.5°C before melting. At the DSC thermogram taken at the same time, an endothermic peak observed in a wide temperature range from 75°C to 135°C is assigned to the crystalline phase transition from the crystal form II to the HT phase, while the sharp endothermic peak at 151°C corresponds to the melting point. The transition occurs in a wide temperature range with a relatively large enthalpy change. Here, the temperature range from 105°C to 135°C is defined as a temperature region B, the temperature upper 135°C is defined a temperature region C. Respectively, they are the temperature before the HT phase appears, the temperature range in which the HT phase increases, and the temperature which a crystals are melting. The endotherm is slightly observed between 75 °C - 105 °C, and this temperature range is defined as a temperature region A.

In Figure 3, the temperature dependences of the L, d_c, the size of the diagonal direction of a crystal lattice of a crystal form II and a HT phase in the lateral dimension of lamellae, integrated peak intensities of $(110)_{II}$, $(100)_{HT}$, and amorphous halo evaluated from WAXD and SAXS data are compared with the DSC thermogram taken simultaneously. In the temperature region A, the crystallite sizes corresponding to the lateral dimension of lamellae were almost constant, the d_c increased slightly, the integrated intensity of the crystal form II gradually decreased, the endotherm observed above 75°C slightly. In the temperature region B, the d_c increased remarkably, the crystallite size of the crystal form II decreased, and the decrease in the integral intensity of the crystal form II and increase of the HT phase and the endotherm corresponding to them are observed. In the region C, the drastic structural change was observed before or after a melting temperature. The L increases, the crystallite size of the crystal form II decreased, the integral intensity of the HT phase decreased, and the intensity of amorphous halo increased drastically.

The phenomenon in the temperature region A was considered as follows; the intermediate phase among the crystal form II and the amorphous phase changes to the amorphous at lower temperature in which the HT phases appeared. In the temperature region B, chains near the surface of lamellae would order and a part of lamellae change into amorphous due to the activated thermal motion of the molecular chains along the chain axis. In the temperature region C in which the thermal motion of the molecular chains is activated along the chain axis, nano-scale crystals would be melting and meso-scale lamellar morphology would change.

4. Conclusions

In the present paper we have described the relationship between morphological change and crystal structure change occurring during the heating process of melt-quenched VDF 82% copolymer sample by using the newly-built-up DSC/WAXD/SAXS simultaneous system with highly brilliant synchrotron X-ray beam of SPring-8. In the temperature from ca. 75°C to ca. 105°C (temperature region A), lamellar staking distance and lamellar thickness evaluated in SAXS data analyses slightly increased though the crystallite size was almost constant. The integrated peak intensities of (110) reflections for crystal form II gradually decreased and scattering from the amorphous region increased instead, which suggested that lamellae in crystal form II became partially disordered before the phase transition.

Acknowledgements. The authors thank Dr. Ohta, Dr. Inoue and Dr. Yagi, Japan Synchrotron Radiation Institute/SPring-8 for their helpful discussion about the DSC measurements. The DSC/WAXD/SAXS simultaneous measurements were performed at a synchrotron radiation system SPring-8 with approved numbers 2006A1128 and 2006B1100. They thank also Daikin Kogyo Co. Ltd., Japan for their kind supply of VDF 82% sample.

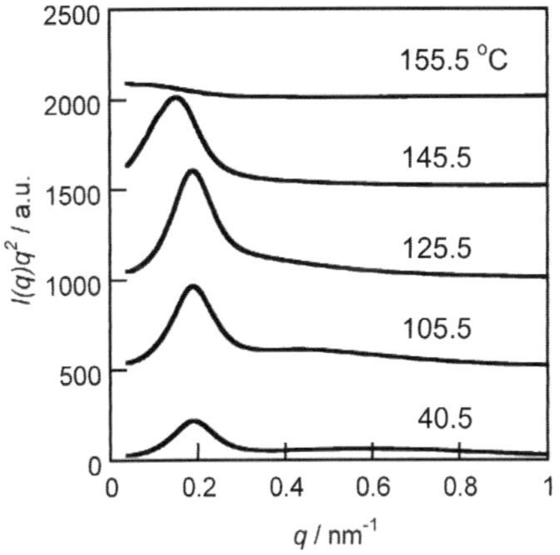

Figure 1. Temperature dependence of SAXS profiles corrected for the Lorentz factor measured for the melt-quenched VDF 82% sample.

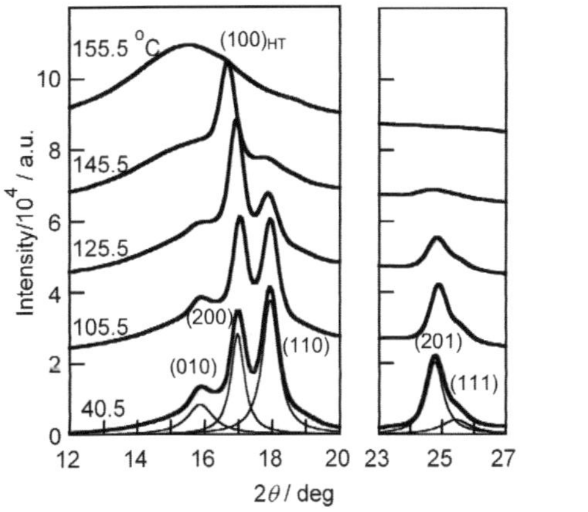

Figure 2. Temperature dependence of WAXD profiles measured for VDF 82% sample. The reflections $(010)_{\mathrm{II}}$, $(200)_{\mathrm{II}}$, $(110)_{\mathrm{II}}$, $(201)_{\mathrm{II}}$, $(111)_{\mathrm{II}}$ are for the crystal form II. A peak observed near 17° observed at 145°C is the reflection (100) of the HT phase.

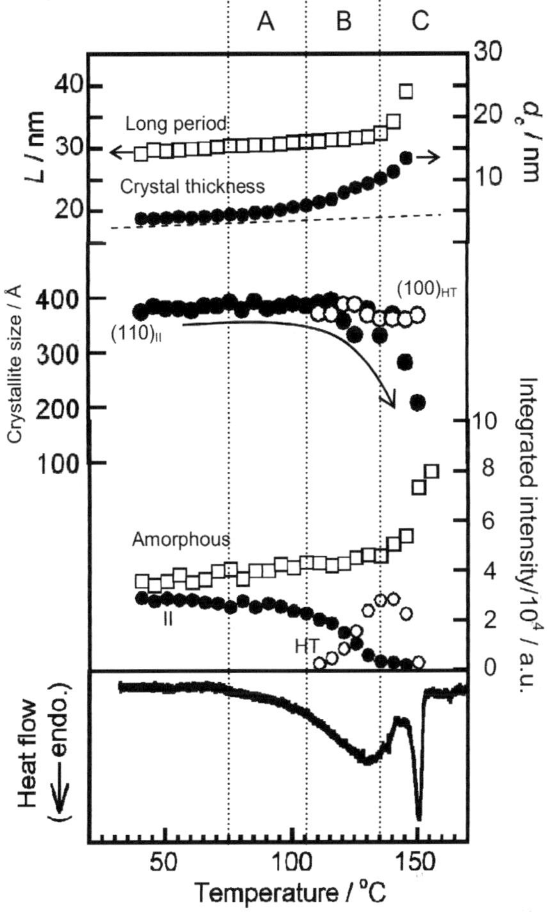

Figure 3. Temperature dependences of L and d_c estimated from the SAXS data, the crystallite size estimated from $(110)_{\mathrm{II}}$, $(100)_{\mathrm{HT}}$, WAXD integrated intensities of $(110)_{\mathrm{II}}$, $(100)_{\mathrm{HT}}$ and amorphous halo, and DSC thermogram measured for VDF 82% sample in the heating process.

References

[1] Tashiro K, In Nalwa, HS editor, 1995 *Ferroelectric polymers* New York: Marcel Dekker: chapter 2

[2] Lovinger A J 1985 *Jpn. J. Appl. Phys. Suppl.* **24** 18

[3] Tashiro M, Kobayashi M 1985 *Polymer Prepr. Jpn.* **34** 2241

[4] Masunaga H, Sasaki S, Tashiro K, Hanesaka M, Takata M, Inoue K, Ohta N, and Yagi 1997 *Polym. J.* **39** 1281

[5] Stroble, G. 1996 *The Physics of Polymers*, Springer 408

[6] Hasegawa R, Takahashi Y, Chatani Y, Tadokoro H 1972 *Polymer J.* **3** 600

[7] Tashiro K, Takano K, Kobayashi M, Chatani Y, Tadokoro H 1984 *Polymer* **25** 195

[8] Tashiro K, Tanaka R 2006 *Polymer* **47** 5433

Journal of Physics: Conference Series **184** (2009) 012015 doi:10.1088/1742-6596/184/1/012015

Crystallization Behavior of Polyethylene on Silicon Wafers in Solution Casting Processes Traced by Time-resolved Measurements of Synchrotron Grazing-Incidence Small-angle and Wide-angle X-ray Scattering

S. Sasaki[1*], H. Masunaga[1], K. Itou[2], K. Tashiro[3], H. Okuda[4], A. Takahara[5], M. Takata[1, 2, 6]

[1]Japan Synchrotron Radiation Research Institute (JASRI) / SPring-8, Hyogo 679-5198, Japan; [2]The RIKEN Harima Institute / SPring-8, Hyogo 679-5198, Japan; [3]Graduate School of Engineering, Toyota Technological Institute, Nagoya 468-8511, Japan; [4]Graduate School of Engineering, Kyoto University, Kyoto 606-8501, Japan; [5]Institute for Materials Chemistry and Engineering, Kyushu University, Fukuoka 819-0395, Japan; [6]Graduate School of Frontier Science, The University of Tokyo, Chiba 277-8561, Japan.

Corresponding author's E-mail: sono@spring8.or.jp

Abstract. Crystallization behavior of polyethylene (PE) on silicon wafers in solution casting processes has been successfully traced by time-resolved grazing-incidence small-angle and wide-angle X-ray scattering (GISWAXS) measurements utilizing synchrotron radiation. A *p*-xylene solution of PE kept at ca. 343 K was dropped on a silicon wafer at ca. 298 K. While the *p*-xylene evaporated naturally from the dropped solution sample, PE chains crystallized to be a thin film. Raman spectral measurements were performed simultaneously with the GISWAXS measurements to evaluate quantitatively the *p*-xylene the dropped solution contained. Grazing-incidence wide-angle X-ray scattering (GIWAXS) patterns indicated nucleation and crystal growth in the dropped solution and the following as-cast film. GIWAXS and Raman spectral data revealed that crystallization of PE was enhanced after complete evaporation of the *p*-xylene from the dropped solution. The [110] and [200] directions of the orthorhombic PE crystal became relatively parallel to the wafer surface with time, which implied that the flat-on lamellae with respect to the wafer surface were mainly formed in the as-cast film. On the other hand, grazing-incidence small-angle X-ray scattering (GISAXS) patterns implied formation of isolated lamellae in the dropped solution. The lamellae and amorphous might alternatively be stacked in the preferred direction perpendicular to the wafer surface. The synchrotron GISWAXS experimental method could be applied for kinetic study on hierarchical structure of polymer thin films.

1. Introduction

Polymer films have widely been utilized as materials and parts of components in the industrial field. The solution casting method is one of the common film molding methods for polymers. However, hierarchical structure development of polymers during the solution casting hasn't been revealed on the meso-to-nano scale. In the case of solution casting with crystalline polymers, polymer chains crystallize as the concentration of polymers increases in the solution due to evaporation of the solvent from it. Therefore, the try-and-error method is necessary to find the best casting condition for control of thickness and morphology of the as-cast films. Crystallization kinetics of polymers and the role of the solvent in the solution casting process have to be clarified to understand relationship between the hierarchical structure and physical property of polymers in the as-cast films.

The grazing-incidence small-angle X-ray scattering (GISAXS) measurements using synchrotron radiation was powerful for hierarchical structure analyses of polymer thin films.[1-7] Ultra-bright and highly directional X-ray beams generated in the top-up mode in SPring-8 are of great advantage to kinetic study of organic compounds and polymeric materials. We have established a simultaneous measurement system of the grazing-incidence small-angle and wide-angle X-ray scattering (GISWAXS) with two-dimensional (2D) detectors for polymer thin films at the BL40B2 (JASRI Structural Biology II, Bending Magnet) in SPring-8.[5-7] A time-resolved experimental system of GISWAXS had been expected as a new tool for the most advanced research and development of polymer thin films.

The purpose of this study is to clarify crystallization behavior of polyethylene (PE) on silicon wafers in the solution casting processes by time-resolved synchrotron GISWAXS measurements. A simultaneous experimental method of GISWAXS and Raman spectra has been established to investigate the effect of natural evaporation of the *p*-xylene solvent on hierarchical structure development of PE.

2. Experimental Section

A sample used in this study was commercially available polyethylene (PE)(P1979-E, Polymer Source Inc., Mn=12500, Mw/Mn=1.05) as a typical example of crystalline polymers. A ca. 1.0 wt % PE was dissolved in the *p*-xylene at ca. 403 K under N_2 atmosphere. The obtained solution was kept at ca. 343 K at which PE chains still dissolved in the *p*-xylene. As a sample for GISWAXS and Raman spectral measurements, a small amount of the solution was dropped on the native oxide covered surface of a silicon wafer (dimension / mm^3: ca. 2.0 (l) × 5.0 (w) × 0.7 (t)) at 298 K. The *p*-xylene was evaporated naturally from the dropped solution under atmospheric condition. Instead, PE chains were gradually crystallized there. After a short time, a dry thin film of PE was formed on the wafer. In the solution casting process, t was defined the time passed after a small amount of the *p*-xylene solution of PE was dropped on a silicon wafer.

Crystallization behavior of PE in the above-mentioned solution casting process was traced by time-resolved measurements of GISWAXS and Raman spectra. To utilize the high brilliance and highly parallel synchrotron X-rays as the incident beams is effective in detecting scattering from a small sample. GISWAXS measurements were carried out at RIKEN Structural Biology Beamline I (BL45XU, Undulator) in SPring-8 (Hyogo, Japan) to investigate hierarchical structure of PE thin films on the nano and meso scales. Figure 1 shows experimental geometry of GISWAXS and Raman spectral measurements. The components of the scattering vector, **q**, parallel and perpendicular to the sample surface were defined as $\mathbf{q}_y = (2\pi/\lambda)\sin(2\theta_f)\cos(\alpha_f)$ and $\mathbf{q}_z = (2\pi/\lambda)(\sin(\alpha_i) + \sin(\alpha_f))$, respectively, for reflected scattering. Here, α_i is the incident angle of X-ray beams, α_f is the exit angle, λ is the wavelength of incident X-ray beams and $2\theta_f$ is the angle between the scattered beam and the plane of incidence. The subscript s and w indicate the grazing-incidence small-angle X-ray scattering (GISAXS) and the grazing-incidence wide-angle X-ray scattering (GIWAXS) geometries, respectively. A charge coupled device (CCD) detector (C4880-10-14A, Hamamastu Photonics K.K.) combined with an imaging intensifier (V5445P, Hamamastu Photonics K. K.) was used for GISAXS measurements, and a flat panel (FP) detector (C9728DK, Hamamatsu Photonics K. K.) for GIWAXS. Meso-scale structural features of the sample on the wafer, such as lamellar stacking distance and lamellar orientation, can be evaluated on the basis of 2D GISAXS patterns. On the other hand, nano-scale structural information on crystal structure and orientation can be obtained by analyzing 2D GIWAXS patterns. The λ of incident X-rays was 0.09 nm and the sample-to-detector distances were ca. 2290

Journal of Physics: Conference Series **184** (2009) 012015 doi:10.1088/1742-6596/184/1/012015

mm for GISAXS and ca. 70 mm for GIWAXS. A 2100 mm-length vacuum path was utilized between a sample cell and the detector for GISAXS measurements. Scattering patterns from the sample were measured at the α_i of 0.10 deg., which were lower than the critical angle of total external reflection at silicon surface. Therefore, the incident X-rays go through the dropped solution or the as-cast PE film and reflect on the silicon wafer surface. The data collection time per scattering pattern was 0.5 s and the interval time between the two sequent patterns was 1.5 s. The dry chicken collagen fibers and the powder CeO_2 (National Institute of Standards and Technology, U.S.A.) were used for the standard samples for GISAXS and GIWAXS measurements, respectively. The detectable **q**-range of GISAXS was ca. 0.09 nm^{-1} ~ 2.1 nm^{-1} and that of GIWAXS was ca. 10 nm^{-1} ~ 32 nm^{-1}. Raman spectral measurements were carried out with a mobile Raman spectrum unit (Spectra View 2000, Rambda Vision Inc.) to evaluate quantitatively the *p*-xylene the dropped solution contained. The wavelength of the built-in laser of the unit was 785 nm.

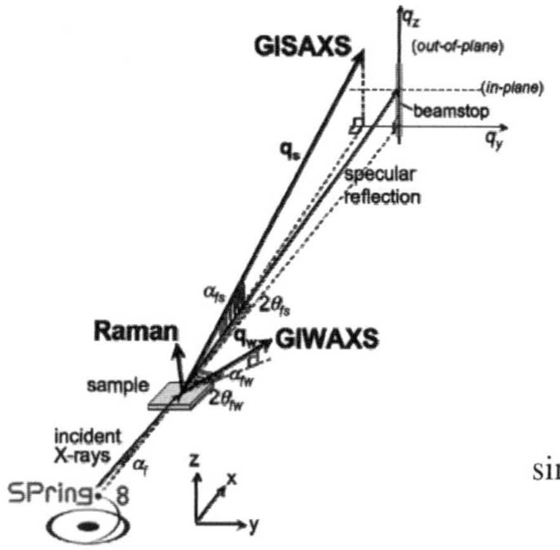

Figure 1. Schematic geometry of simultaneous measurements of GISWAXS and Raman spectra in SPring-8

3. Results and Discussion

Crystallization behavior of PE in the solution casting process was traced by time-resolved simultaneous measurements of GISAXS, GIWAXS and Raman spectrum. The time (t) passed after dropping the *p*-xylene solution of PE on a silicon wafer was recorded. Figure 2 shows time-resolved 2D GISWAXS patterns of the *p*-xylene solution of PE dropped on a silicon wafer measured at t = 20, 120, 126, 128, 130, 132, 134 and 180 s. The 2D GIWAXS patterns indicated that typical isotropic

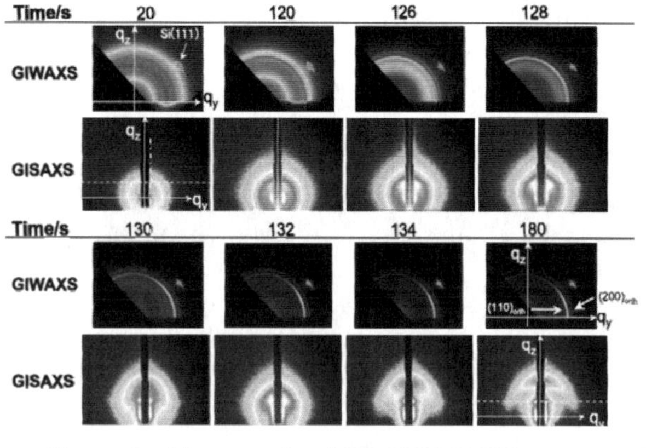

Figure 2. Time-resolved 2D GISWAXS patterns of the *p*-xylene solution of PE dropped on a silicon wafer measured at t = 20, 120, 126, 128, 130, 132, 134 and 180 s.

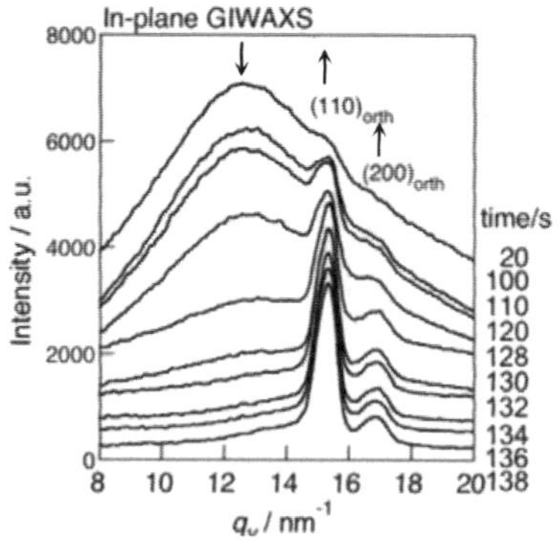

Figure 3. Time dependence of *in-plane* GIWAXS patterns in the solution casting process shown in Figure 2.

scattering from the dropped solution changed gradually to the (110) and (200) reflections from the oriented orthorhombic PE crystals by way of the Debye-Sherrar ring from unoriented those. On the other hand, the 2D GISAXS patterns showed that isotropic scattering from the solution changed gradually to anisotropic one. A broad peak appeared in the *out-of plane* direction on the $\mathbf{q}_z$ axis around 120 s and shifted to the higher $\mathbf{q}_z$ range increasing in intensity until 134 s.

Figure 3 shows time dependence of *in-plane* GIWAXS patterns in the solution casting process shown in Figure 2. The *in-plane* line in the GIWAXS patterns in Figure 2 was parallel to and a few pixels-upper from the $\mathbf{q}_y$ axis. The (110) and (200) reflections from the orthorhombic PE crystals were detected in the $\mathbf{q}_y$ range of 14 ~ 18 nm^{-1} from the begging of the measurements. These reflections increased in intensity with time. And then, broad scattering from the solution and the amorphous PE measured in the $\mathbf{q}_y$ range of 8 ~ 20 nm^{-1} decreased in intensity instead. The profiles didn't change in intensity after 132 s, which suggested that the solvent evaporation and nano-scale crystallization was almost finished around 132 s.

Figure 4 shows Raman spectral profiles in Raman shifts of 750 ~ 1100 cm^{-1} measured in the solution casting process in Figure 2. Strong peaks at ca. 1000 cm^{-1} characteristic of the *p*-xylene decreased gradually in intensity with time. Those spectral peaks completely disappeared around 100 s. Figure 5 shows comparison of peak intensity of the (110) reflection in Figure 3 with that of the Raman spectra at ca. 1000 cm^{-1} in Figure 4. It clearly indicates that the *p*-xylene evaporated completely from the dropped solution around 100 s. Instead, the (110) reflection increased drastically in intensity. It should be mentioned that crystallization of PE was enhanced after complete evaporation of the *p*-xylele. Due to the latent heat of vaporization of the *p*-xylele, the sample might be cooled from ca. 343 K to ca. 298 K during the GISWAXS measurements. Therefore, it was considered that PE chains in the amorphous state could crystallize in such cooled sample after complete evaporation of the *p*-xylele.

Figure 6a and 6b show time dependence of the azimuthal intensity profiles of the (110) and (200) reflections in Figure 3. The azimuthal angles on the *in-plane* line parallel to the $\mathbf{q}_y$ axis and on the $\mathbf{q}_z$ axis were defined as 0 deg. and 90 deg., respectively. The intensity of these azimuthal profiles near 0 deg. increased gradually with time. As time passed, broad peaks around 0 deg. in these figures increased in intensity. This means that the [110] and [200] directions of the orthorhombic PE crystal became relatively parallel to the wafer surface. In other words, it was found that the *a* and *b* axes of the orthorhombic PE crystal gradually oriented parallel to the wafer surface with time. At the same time, the *c* axis (// the chain axis) became perpendicular to the wafer surface. The flat-on lamellae with respect to the wafer surface might mainly be formed in the end of the solution casting process. Figure 7 shows time dependence of the *out-of-plane* GISAXS profiles at $\mathbf{q}_y$ = ca. 0.11 nm^{-1} (along the broken line parallel to the $\mathbf{q}_z$ axis) in Figure 2. At the beginning before 128 s, a board peak was observed near $\mathbf{q}_y$ = 0, which indicated formation of isolated lamellae in the dropped solution. From 128 s to 132 s, a shoulder peak appeared, shifted to the higher $\mathbf{q}_z$ of ca. 0.5 nm^{-1}, and increased in intensity. After 132 s, the profile showed little change in intensity. This indicated that the flat-on lamellae and amorphous might alternatively be stacked preferentially in the perpendicular direction to the wafer surface after complete evaporation of the *p*-xylele. The long

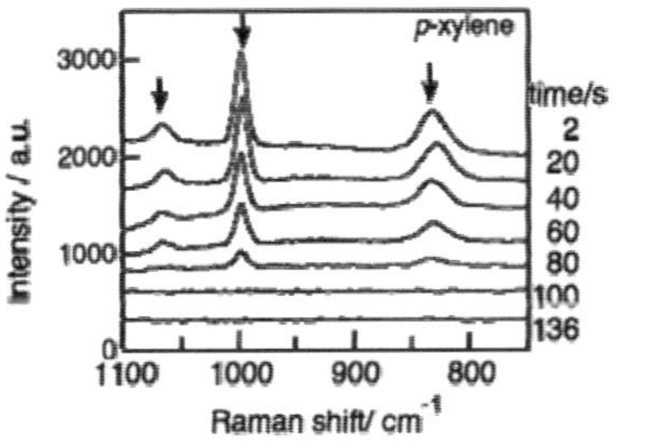

Figure 4. Raman spectral profiles in Raman shifts of 750 – 1100 cm^{-1} measured in the solution casting process in Figure 2.

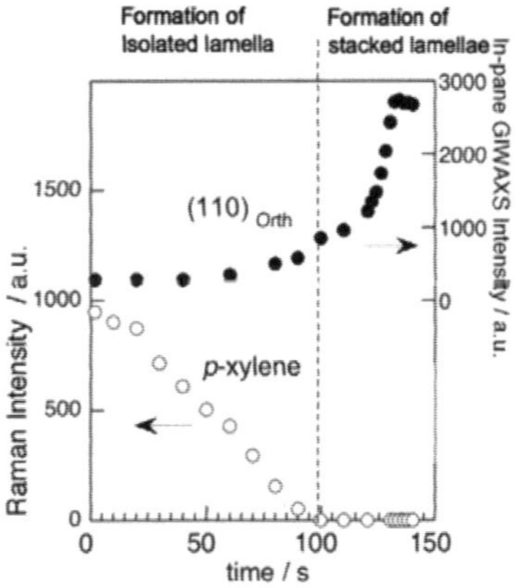

Figure 5. Comparison of peak intensity of the (110) reflection in Figure 3 with that of the Raman spectra at ca. 1000 cm^{-1} in Figure 4.

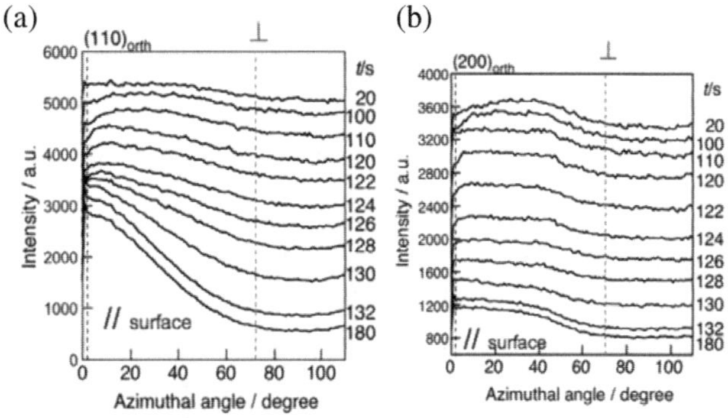

Figure 6. Time dependence of the azimuthal intensity profiles of the (110) and (200) reflections in Figure 3 (a and b).

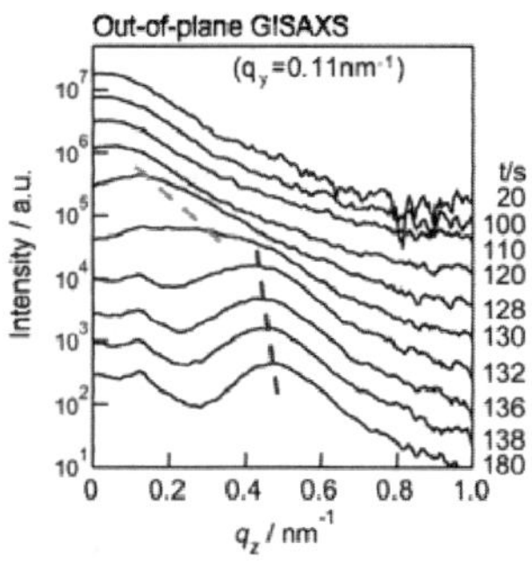

Figure 7. Time dependence of the *out-of-plane* GISAXS profiles in the solution casting process in Figure 2.

period of this lamellar stacking structure estimated from the peak position was ca. 12 nm for the as-cast film.

The GISWAXS and Raman spectral data suggested that the crystallization behavior of PE chains in the solution casting process could be explained as the following. (1) Nucleation of the orthorhombic PE crystals occurred in the dropped solution during evaporation of the *p*-xylene from it. (2) The nucleuses grew to be isolated lamellar crystals. (3) After complete evaporation of the *p*-xylene from the dropped solution, PE chains in the amorphous state crystallized rapidly. (4) The lamellae oriented gradually flat-on with respect to the wafer surface. (5) Alternative stacking of the flat-on lamellae and amorphous occurs preferentially in the perpendicular direction to the wafer surface. (6) Lamellar stacking structure with the long period of ca. 12 nm was established in the as-cast film.

4. Conclusions

In this study, the time-resolved and simultaneous experimental method of synchrotron GISWAXS and Raman spectra has been established with 2D detectors at the BL45XU in SPring-8. On the basis of this method, nucleation, formation, alignment and orientation of PE crystals on silicon wafer substrates could successfully be traced in the solution casting processes on the molecular and lamellar scales. It was found that the *p*-xylene might play a very important role especially in crystal nucleation. After complete evaporation of the *p*-xylene from the dropped solution, crystallization of PE chains in the amorphous state and formation of lamellar stacking structure was enhanced. Time-resolved synchrotron GISWAXS experimental method can be applied to a various kinds of kinetic researches in the thin-film, surface, and interface science of polymeric materials from the academic and industrial viewpoints.

Acknowledgement

The synchrotron radiation experiments were performed at RIKEN Structural Biology Beamline I (BL45XU) in SPring-8 with the approval of RIKEN (Proposal No. 20070150).

References

[1] P. Busch, M. Rauscher, D.-M. Smilgies, D. Posselt, and C. M. Papadakis, *J. Appl. Cryst.*, **39**, 433-442 (2006) .

[2] P. Busch, S. Krishnan, M. Paik, G. E. S. Toombes, D.-M. Smilgies, S. M. Gruner, and C. K. Ober, *Macromolecules,* **40**, 81-89 (2007).

[3] G. E. Stein, E. J. Kramer, X. Li, and J. Wang, *Macromolecules*, **40**, 2453-2460 (2007).

[4] J. Yoon, K. S. Jin, H. C. Kim, G. Kim, K. Heo, S. Jin, J. Kim, K.-W. Kim, and M. Ree, *J. Appl. Cryst.*, **40**, 476–488 (2007).

[5] S. Sasaki, H. Masunaga, H. Tajiri, K. Inoue, H. Okuda, H. Noma, K. Honda, A. Takahara, and M. Takata, *J. Appl. Cryst.,* **40**, s642-s644 (2007).

[6] S. Sasaki, H. Masunaga, H. Tajiri, H. Okuda, K. Inoue, A. Takahara and M. Takata, *Trans. Mater. Res. Soc. Jpn.*, **32**(1), 193-197 (2007).

[7] S. Sasaki, H. Okuda and T. Takata, *SPring-8 Research Frontiers*, Japan Synchrotron Research Institute/SPring-8, 2007, 102-103 (2007).

Journal of Physics: Conference Series **184** (2009) 012016 doi:10.1088/1742-6596/184/1/012016

Brittle, ductile, paste-like behaviors and distinct necking of double network gels with enhanced heterogeneity

Yasunori Kawauchi[1], Yoshimi Tanaka[1], Hidemitsu Furukawa[1], Takayuki Kurokawa[1], Tasuku Nakajima[1], Yoshihito Osada[2] and Jian Ping Gong[1]

[1]Laboratory of Soft and Wet Matter, Division of Biological Science, Graduate School of Science, Hokkaido University, Sapporo 060-0810, Japan

[2]RIKEN, 2-1, Hirosawa, Wako, Saitama 351-0198, Japan

gong@mail.sci.hokudai.ac.jp

Abstract. We propose a simple method to obtain double network (DN) hydrogels containing 85-90wt% water with various kinds of mechanical performance by taking advantage of the heterogeneity of the first network: By synthesizing the first network in a poor solvent that enhancing the heterogeneity of the network, the DN gel exhibits rich mechanical behavior: from brittle, ductile (necking), to paste-like. For a proper heterogeneity, a distinct necking deformation with extensibility of about 20 times is observed for post-necking gels. The necking gels have a high initial modulus (0.1-0.5 MPa) and high mechanical toughness (nominal fracture stress: 1 MPa), both values are similar to that of living tissues.

1. Introduction

Living tissues, containing 50-80 wt% water, are soft and wet materials with an excellent mechanical property: They have the elasticity in a range of several 100 kPa to MPa, and sustain deformations as large as 100% and failure stresses of 1-10MPa [1]. Material scientists are facing challenges in providing synthetic soft tissues using hydrogels, the same soft and wet materials. Common hydrogels are usually mechanically weak due to 1) high water contents; and 2) heterogeneity in the cross-link structure either formed by covalent cross-links (chemical gel) or attractive interactions (physical gel). High water content leads to low elasticity and heterogeneity concentrates the tension, bringing about failure of the hydrogels at small stresses.

To reduce the high water content (usually higher than 95 wt% for a common hydrophilic gel), a common method is to increase the cross-linking density, this in turn, enhances the heterogeneity and the gel becomes brittle.

One method to avoid the concentrated stress caused by heterogeneity is with a sliding cross-linking structure (sliding ring gel): polymer chains with bulky end groups are interlocked by figure-of-either cross-links. Hence, these cross-links can pass along the polymer chains freely to equalize the tension of the threading polymer chains and sustain large deformation [2].

Journal of Physics: Conference Series **184** (2009) 012016 doi:10.1088/1742-6596/184/1/012016

Here, we propose an alternative method to obtain a tough gel containing 85-90wt% water by taking advantage of the heterogeneity of double network (DN) gels: We demonstrate how the mechanical toughness be improved by enhancing the heterogeneity in the double network (DN) gels. By synthesizing the first network in a poor solvent that enhancing the heterogeneity of the network, the DN gel exhibits rich mechanical behavior: from brittle, ductile (necking), to paste-like. For a proper heterogeneity, a distinct necking deformation with extensibility of about 20 times is observed for post necking gels. The necking gels have a high initial modulus (0.1-0.5 MPa) and high mechanical toughness (nominal fracture stress: 1 MPa), both values are similar to that of living tissues [1].

2. Experiment

The synthesis procedure of the DN gel, consists of densely cross-linked polyelectrolyte, poly(2-acrylamido-2-methylpropanesulfonic acid)(PAMPS), as the first network and loosely cross-linked neutral polymer, polyacrylamide(PAAm), as the second network, is identical to previous studies [3-6], except that the polymerization (gelation) reaction of the first network (PAMPS) is carried out in special reaction solvent, mixtures of water and 1,4-dioxane (details for sample preparation are given in the next paragraph). Below the volume fraction of 1,4-dioxane in the reaction solvents is denoted by ϕ. The solvents can dissolve AMPS monomer (and cross-linker) but the existence of 1,4-dioxane induces weak segregation of the PAMPS during the polymerization, leaving enhanced heterogeneity in the first network.

2-Acrylamido-2-methylpropanesulfonic acid (AMPS; Tokyo Kasei Co., Ltd.) was used as received. Acrylamide (AAm; Junsei Chemical Co. Ltd) was recrystallized from chloroform. N,N'-Methylenebis-(acrylamide) (MBAA; Tokyo Kasei Co., Ltd.), a cross-linking agent for both AMPS and AAm gels, was recrystallized from ethanol. 2-oxoglutaric acid (Wako Pure Chemical Industries, Ltd.), a radical initiator for the gelation reactions, was used as received. The PAMPS gels are synthesized by free-radical polymerization. 1 M AMPS, 4 mol % of MBAA, 0.1 mol % of 2-oxoglutaric acid (the molar percentages, 4 and 0.1 mol %, are determined with respective to the AMPS monomer) were dissolved into the mixture solvents. Photo-initiated polymerization was carried out under argon atmosphere with an ultraviolet (UV) lamp for 9 h (The distance between the lamp and the sample mold was about 200 mm.). The DN gels are synthesized through a two-step sequential free-radical polymerization. The first step is described above. In the second step, after the gelation was completed, the platy PAMPS gel was immersed into 2 M AAm solution containing 0.1 mol % of 2-oxoglutaric acid and prescribed amount of MBAA (0 ~ 1mol %) for at least 2 days until the equilibrium was reached. By irradiation with the UV lamp for 8 h, the second network was subsequently synthesized in the presence of the first network. Samples were swollen in water for 1 week. The main material parameter in this study is the volume fraction ϕ of the dioxane. In addition, we changed the cross-linker (MBAA) density in the second PAAm network to obtain data in Fig. 4. The final water content of the DN gel is 85-90wt%.

The internal structure of the PAMPS gels in water was observed by confocal laser scanning microscope (BX 50, Olympus Co. The laser wavelength was 568nm) with 60x ojective lens (N.A. = 1.4) in refractive mode without any fluorescent dye, where the special resolution was estimated about 200nm in an ideal condition.

Dynamic light scattering measurements of the PAMPS gels were performed with a scanning microscopic light scattering system [7]. Dynamic light scattering experiments were carried out on a compact goniometer at respective angles of 40, 60, 90, 125°. A 22 mW He-Ne laser (the wavelength in a vacuum: $\lambda = 632.8$nm) was used as the beam through a cylindrical sample tube with a 10 mm outer diameter. Before the measurement, the PAMPS gels were immersed in 1mM NaCl aqueous solutions for two days, in order to modify the A typical measuring time at a position was 90 s. The temperature of the sample was kept at 30.0 °C. 31 points per sample were measured with the stepping motor moving the sample holder vertically in 30μm steps. The dynamic component of the ensemble-averaged correlation function was rigorously determined from the 31 data points and was transformed into a relaxation-time distribution function $P(\tau_R)$ by using numerical inversed Laplace transformation method [7], where τ_R is the relaxation time. Then the two relaxation modes were quantitatively characterized. It is very similar to the results of previous studies about the two modes in the DN gels [8-10] In the present study, we simply estimate the size of characteristic fluctuation from the behavior of both the faster and slower modes ξ, based on the Stokes-Einstein equation $\xi = k_B T|\mathbf{q}|^2 \tau_R / 6\pi\eta$, where k_B is the Boltzmann constant, T is the absolute temperature, $|\mathbf{q}|$ is the amplitude of scattering vector, and η is the viscosity of solvent.

The tensile mechanical properties were measured with a commercial test machine (Tensilon RTC-1150A, Orientec Co.). The gels were cut into a dumbbell shape standardized as the JIS-K6251-7 sizes (length: 35 mm; width: 6 mm; thickness: 2-3 mm; gauge length: 12 mm; inner width: 2 mm). The nominal stress was used in the tensile curves. The tensile strain rate was $1.3\times10^{-1}\,s^{-1}$.

3. Result and discussions

Fig. 1a shows appearance of PAMPS gels synthesized at different ϕ. For $\phi \geq 0.6$, the gels were turbid and the turbidity increased with ϕ; the increase in turbidity indicates the existence and growth of heterogeneity on the scale of visible light wavelength. This also suggests that even for $\phi < 0.6$ where the gels were transparent, the heterogeneity gradually increased with ϕ.

The internal structure in gels were observed by a confocal scanning microscope (CLSM) as seen in Figs. 1b and 1c. The inner structure in the PAMPS gels at $\phi = 0.8$ looks more inhomogeneous than that at $\phi = 0$.

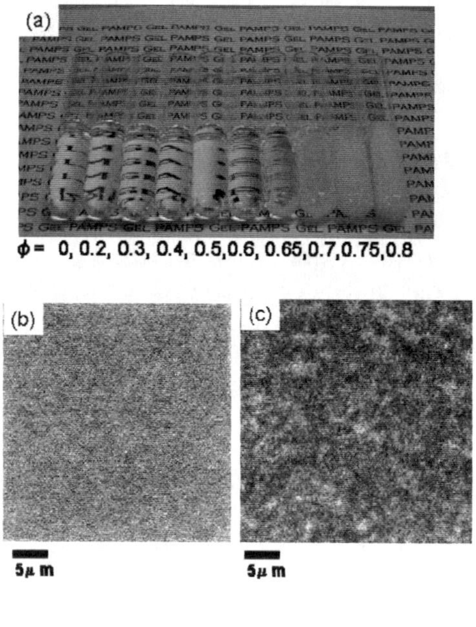

Figure 1. (a): The first PAMPS gels synthesized in water and 1,4-dioxane mixed reaction solvents with different volume fraction of poor solvent, 1,4-dioxane ϕ; (b, c): Confocal laser scanning microscope images of the PAMPS gel at ϕ=0 (a) and ϕ=0.8 (b).

Journal of Physics: Conference Series **184** (2009) 012016 doi:10.1088/1742-6596/184/1/012016

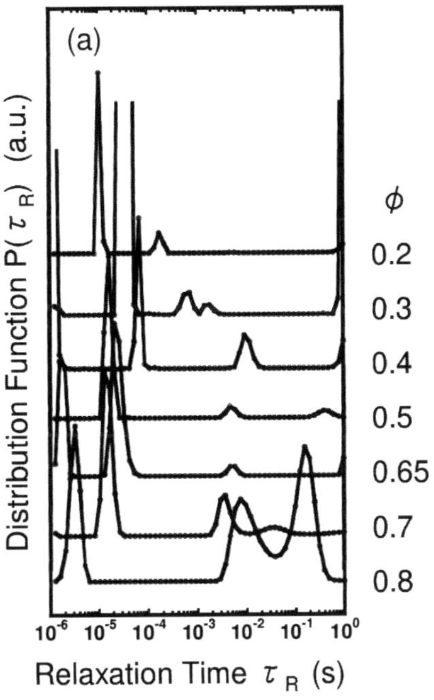

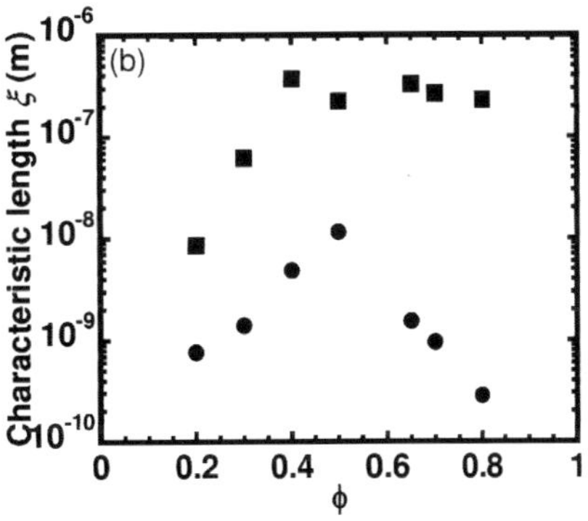

Figure 2. (a) Distribution function of the relaxation spectrum, at the different volume fraction of poor solvent ϕ. (b) Characteristic length estimated from the distribution function (a): (●) the shorter length assigned to the mesh size in the network structure. (■) the longer length assigned to the large size of inhomogeneities due to the gelation with the poor solvent.

In order to characterize the heterogeneous structure in the PAMPS gels more precisely, the dynamic light scattering was performed. Fig. 2a shows the relaxation-time distribution of the DN gels prepared at different fraction of poor solvent ϕ. For all the gels, the fast mode with a sharp peak profile and the slow mode with uni- or bimodal peak profile were observed. Here we consider that the bimodal peak profile we observed is not correct. The true profile of the slow mode is probably made from an unimodal peak with wide distribution. Thus, we estimated the two kinds of characteristic length ξ from the fast and slow modes, as shown in Fig. 2b. We consider that the short length is related to the mesh size in the DN structure and the long length is related to the large-scale inhomogeneity induced by the addition of the poor solvent. For $\phi < 0.6$, both the sizes of the network structure and the large-scale inhomogeneity increase as ϕ increases. On the other hand, for $\phi \geq 0.6$, the size of the network structure decreases but the size of the inhomogeneous structure increases as ϕ increase. The behavior of the inhomogeneous structure is consistent with the behavior of the turbidity, as above mentioned. Also, the behavior of the mesh size is also consistent with the behavior of the swelling weight ratio, as will be described below.

Fig. 3a shows the behavior of the equilibrium weight ratio q of the PAMPS gels used as the first network and the initial Young's modulus E as functions of ϕ. Here, q is the weight ratio of PAMPS gel at swelling equilibrium in water to that after drying. Since the weight of the swollen gel is dominated by water, $1/q$ is approximately proportional to the concentration (g/*l* or mol/ *l*) of PAMPS in gel, and also to that in *the resultant DN gels* [11]. As shown in Fig. 3a, for $\phi < 0.6$, q increases as ϕ increases, and for $\phi \geq 0.6$, q decreases as ϕ increases. This behavior is the similar to the behavior of the mesh size. It implies that the change in the swelling weight ratio comes from the change in the mesh size.

To elucidate the relationship between the q and E, we plot of the same data as the relation of E versus q, in Fig. 3b (ϕ is indicated in the plot.) For $\phi \leq 0.65$, E decreased with q; this behavior is

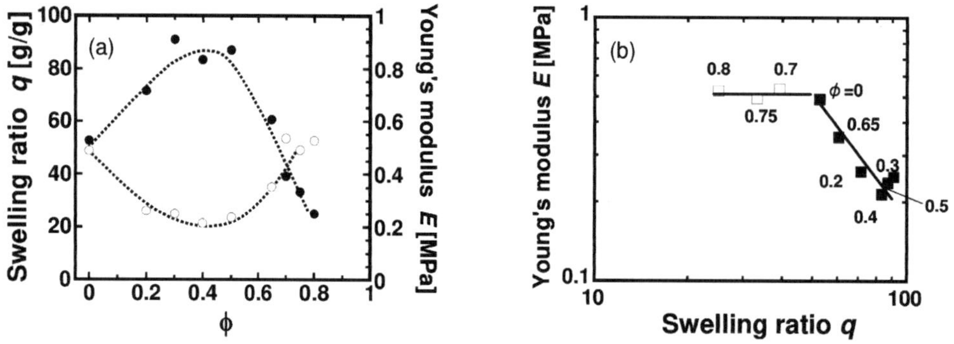

Figure 3. (a): Dependence of swelling ratio q (●) and Young's modulus E (○) on the volume fraction of poor solvent ϕ, where q is defined as the ratio of the weights of the PAMPS gels at swelling equilibrium in water and at dried state; (b): The corresponding plot of E versus q.

quite usual for polyelectrolyte gel, because E is roughly proportional to osmotic pressure of counter ions from PAMPS. On the other hand, E is almost independent of q for $q \geq 0.7$. This indicates that for the turbid gels, the counter ions are confined in PAMPS rich domains, which reduces the activity of the counter-ions and the apparent (whole) swelling volume of the gels is not simply determined by the counter-ion concentration.

Fig. 4a gives a comparison of loading curves of the DN gels with different ϕ (and uncross-linked second network, tensile strain rate is 1.3×10^{-1} s^{-1}), and the stretching process of the sample DN gel with $\phi = 0.2$ (photos A-G). The initial (low strain) behaviors of the loading curves are similar with each other. On the other hand, the yielding and the fracture behaviors strongly depend on ϕ. For $0.2 \leq \phi \leq 0.4$, the DN gels showed lower yielding stresses (around 0.5 MPa) and fairly higher fracture strain (larger than 15), compared with the gels with $\phi = 0$ [12]. At $\phi = 0.5$, the gel underwent necking, however it broke during necking. Above $\phi = 0.5$, the gels got pasty and sticky, and broke quite easily without necking. As seen in photographs (and data points) C, D, and E, a plateau region was created in the loading curve during the necking propagation (the arrows in the photos shows a boundary of the necking zone). After the boundary of the neck reached around the clamped parts (F), the stress re-increased with further stretching and the necked gel was uniformly stretched withstanding high extensions (G). The post-necking gels almost returned to the original lengths when the tensile force was removed (although a few percents of permanent strains remained). Figs. 4b and 4c show the yielding stress (and strain) and the fracture stress (and strain) as a function of ϕ, respectively. The yielding stress monotonously decreased with ϕ. The yielding strain showed similar behavior, but the dependence was weaker. On the other hand, the fracture stress (strain) had a peak around $\phi = 0.2$ ($\phi = 0.3$). The peak corresponds to the result that only the DN gels with intermediate ϕ can withstand the large extensions in Fig. 4a.

In our previous work, some DN gels with modified first network structure show necking phenomena during tensile tests [6]. We proposed a structural model of the necked gel that the first network breaks into small fragments and they are interconnected by the second PAAm chains. The monotonously decrease of the yielding stress (in Fig. 4b) can be directly understood based on the structural mode: the heterogeneity works as mechanical defects and expedites the onset of catastrophic breakdown of the

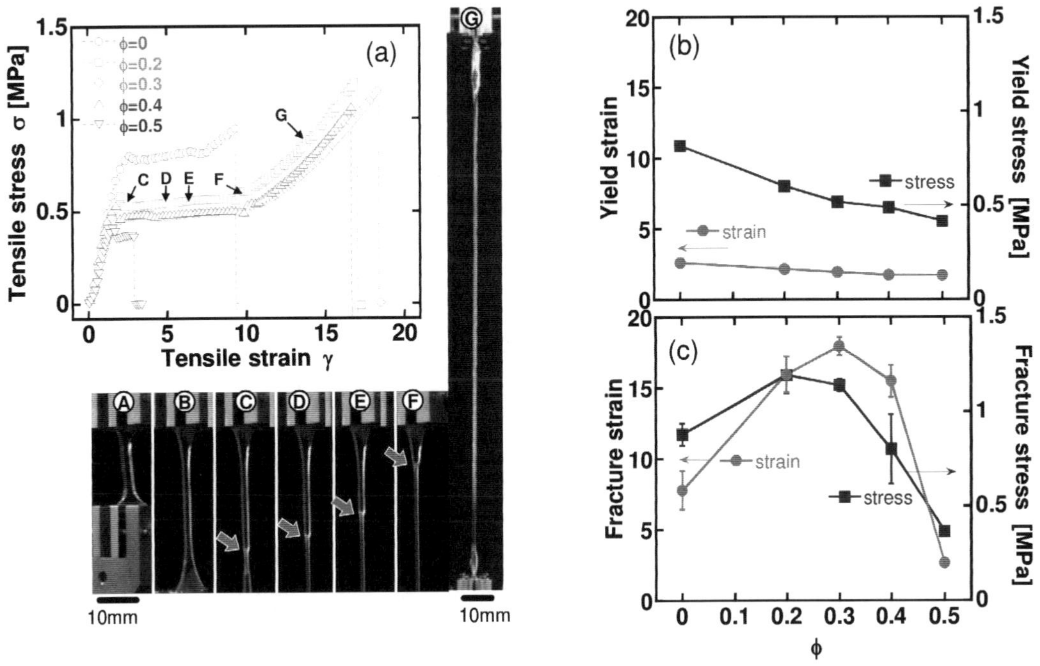

Figure 4. (a): A comparison of tensile curves of the DN gels with different ϕ, and photographs of the necking process for the DN gel with $\phi = 0.2$. The tensile strain rate is $1.3 \times 10^{-1}\,\mathrm{s}^{-1}$. The photos of C-E correspond to the neck propagation. During this period, a plateau was produced in the loading curve. (b): The dependence of the yielding strain (and stress) on ϕ determined from the curves; (c): The dependence of the fracture strain (and stress) on ϕ determined from the curves.

first network. If post-necking gel with high extensibility is free from large scale heterogeneities that causes strong stress concentration, its maximum breaking strain λ_b could be related to the ratio of the average contour length L_c to the average radius R of the partial PAAm chains between two cross-linked points: $\lambda_b \cong L_c / R$. Employing $L_c = aN$ and $R \cong aN^{3/5}$ [13], $\lambda_b \cong N^{2/5}$, here N is the Kuhn number, which is approximately equal to the repeated units of PAAm, between the effective cross-linkages of PAAm. With the observed value of breaking strain, $\gamma_b \approx 17$, and the relation $\lambda_b = \gamma_b + 1$, we have $N \cong 10^3$. This order estimation for N is consistent with an estimation based on our very recent experiments [14]. On the other hand, it is impossible at present to explain why the DN gels with intermediate heterogeneity show the highly extensibility because of lack in detailed information about the structure of the post-necking gels and their fracture mechanism. At least, however, we can interpret the deterioration in the mechanical properties for larger ϕ as follows: when the spatial scale of the heterogeneity is larger than average radius R of PAAm random coils (around a few hundreds nm [5]), the PAAm chains filling void (or PAMPS poor) region behave like in the bulk state, that is, a flowable sol; thus they are incapable of connecting with PAMPS rich domains after onset of breaking-up of the first network.

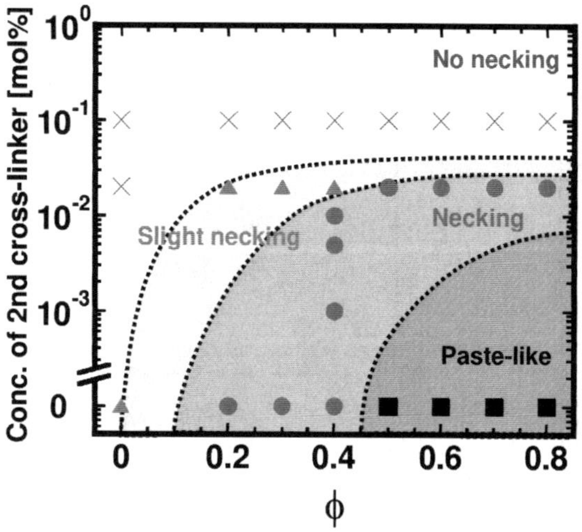

Figure 5. A phase diagram-like representation of tensile behavior of PAMPS/PAAm DN gels with different volume fraction of 1,4-dioxane ϕ for synthesizing the first network, and different cross-linker MBAA density y of the second network.

The above argument can be verified by increase the cross-linker, N,N'-Methylenebis-(acrylamide) (MBAA), in the formation of the second network, PAAm. In Fig. 5, we give a phase diagram-like representation of the tensile behavior of the DN gels. To construct the diagram, we varied the cross-linker concentration y of the second network, MBAA (the concentration is represented in the unit of mol% with respect to AAm monomer), in addition to ϕ. As mentioned above, the tensile behavior of DN gels with $y = 0$ (second network is polymerized without adding cross-linker) changed with ϕ in the following manner: (i) at $\phi = 0$, the gel underwent 'moderate' necking, i.e., the gel showed relatively low extensibility and sometimes broke during neck propagation; (ii) for $0.2 \le \phi \le 0.4$, the gels showed pronounced necking with high extensibility; (iii) for larger ϕ, the gels broke quite easily without necking.

The necking behavior (dependence on ϕ) at $y = 0.02$ shifts to the larger ϕ side compared with the behavior at $y = 0$: for $0.2 \le \phi \le 0.4$ the gels showed the moderate necking, and for $\phi > 0.5$, the pronounced necking occurred. These results show that the necking behavior of the DN gels is determined by a balance between the fragility of the first network and the ductility of the second network.

On the other hand, for the highest value of $y = 0.1$, the necking never occurs for all ϕ; the gels behave like usual chemical gels and suddenly broke during stretching. This behavior is reasonable because at this cross-linker density, the bare second PAAm network itself behaves as a densely chemically cross-linked gel that is brittle-like and that never show necking.

In summary, we have studies the tensile behavior of the DN gels containing the first network synthesized in the mixed solvent of water and 1,4-dioxane. Our results clearly show that using the mixed solvent is a good way to systematically *enhance* the heterogeneity of the first network, and to control the mechanical properties of the resultant DN gels from brittle, ductile, necking, to pasty. The condition to realize the pronounced necking deformation and to obtain highly extensible DN gels are: (1) the second network should be synthesized with a little amount of cross-linker (or even without cross-linker), and (2) it is preferable to introduce a certain degree of heterogeneity in the fist network; with increasing the heterogeneity, the DN gels show lower yielding stress and higher extensibility. (Too strong heterogeneity, however, results in drastic deterioration in mechanical properties for the uncross-linked second network.) The present study proposes a new method to realize flexible and tough hydrogels with an elasticity of several hundred kPa and breaking stress in MPa order, which are comparable to those of living tissues [1].

Acknowledgments
The present work was supported by the Grants-in-Aid for Specially Promoted Research (No. 18002002) from the Ministry of Education, Culture, Sports, Science and Technology (MEXT), Japan.

References
[1] Fung Y C 1993 *Biomechanics, mechanical properties of living tissues* (New York; Tokyo: Springer-Verlag)
[2] Okumura Y, Ito K 2001 *Adv. Mater.* **13** 485
[3] Gong J P, Katsuyama Y, Kurokawa T and Osada Y 2003 *Adv. Mater.* **15** 1155
[4] Tanaka Y, Kuwabara R, Na Y-H, Kurokawa T, Gong J P and Osada Y 2005 *J. Phys. Chem.* B **109** 11559
[5] Tsukeshiba H, Huang M, Na Y-H, Kurokawa T, Kuwabara R, Tanaka Y, Furukawa H, Osada Y and Gong J P 2005 *J. Phys. Chem.* B **109** 16304
[6] Na Y-H, Tanaka Y, Kurokawa T, Furukawa H, Sumiyoshi T, Gong J P and Osada Y 2006 *Macromolecules* **39** 641
[7] Furukawa H, Horie K, Nozaki R, Okada M 2003 *Phys. Rev.* E **68** 031406-1
[8] Huang M, Furukawa H, Tanaka Y, Nakajima T, Osada Y, Gong J P 2007 *Macromolecules* **40** 6658
[9] Na Y-H, Kurokawa T, Katsuyama Y, Tsukeshiba H, Gong J P, Osada Y, Okabe S, Karino T, Shibayama M 2004 *Macromolecules* **37** 5370
[10] Furukawa H, Kuwabara R, Tanaka Y, Kurokawa T, Na Y-H, Osada Y, Gong J P 2008 *Macromolecules* **41** 7173
[11] This statement is based on the following experimental facts: (i) the equilibrium volume of the PAMPS gels in AAm monomer solutions for the second network, say V_{AAm}, is very close to that in pure water, V_{wat}; (ii) after the formation of the second network, the volume of the resultant DN gels, V_{DN}, is hardly changed from V_{AAm} even after swelling equilibrium in water is accomplished. That is, $V_{\mathrm{DN}} \approx V_{\mathrm{mon}} \approx V_{\mathrm{wat}}$.
[12] It should be commented here that even the original tough DN gel, the one with $\phi = 0$, shows the necking phenomena (although the extension is not so high). The authors had not realized this point when they wrote [1] and [7], because they mainly employed the compression test to characterize the mechanical strength then and the necking did not occur in the compression test.
[13] P G de Gennes 1979 *Scaling Concept in Polymer Physics* (New York: Cornell University Press)
[14] Unpublished data. In [5, 6], we thought that the effective cross-linking of the second network was mainly due to the entanglement between the first and second components. Our very recent experiments, however, show (i) the second PAAm chains are bonded to the first network at the residual functions of the cross-linker in the first network; and (ii) this "graft cross-linking" plays an important role in mechanical properties of DN gels.
[15] Tominaga T, Tirumala V R, Lin E K, Gong J P, Furukawa H, Osada Y, Wu W L 2007 *Polymer* **48** 7449
[16] Tominaga T, Tirumala V R, Lin E K, Gong J P, Wu W L 2008 *J. Phys. Chem.* **112** 3903

Journal of Physics: Conference Series **184** (2009) 012017 doi:10.1088/1742-6596/184/1/012017

Living Anionic Polymerization of Styrenes Containing Adamantyl Skeletons

Shingo Kobayashi, Hiroshi Kataoka, and Takashi Ishizone*

Department of Organic and Polymeric Materials, Tokyo Institute of Technology,

2-12-1-H-119, Ohokayama, Meguro-ku, Tokyo 152-8552 Japan

E-mail: tishizon@polymer.titech.ac.jp

Abstract Anionic polymerizations of a series of *para*-substituted styrene derivatives containing adamantyl groups, 4-(1-adamantyl)styrene (**1**), 3-(4-vinylphenyl)-1,1'-biadamantane (**2**), 4-(1-adamantyl)-α-methylstyrene (**3**), 4-(1-adamantyloxy)styrene (**4**), 1-adamantyl 4-vinylbenzoate (**5**), and *N*-(1-adamantyl)-*N*-(4-vinylbenzylidene)amine (**6**), were carried out in THF at –78 °C. The polymerizations of **1**-**6** quantitatively proceeded to afford the well-defined polymers possessing the predicted molecular weights and the narrow molecular weight distributions (M_w/M_n = 1.03-1.10). The stability of the propagating chain ends was confirmed by the sequential copolymerizations of **1**-**6** with various monomers. The resulting poly(**1**), poly(**2**), poly(**3**), poly(**4**), poly(**5**), and poly(**6**) showed remarkably high glass transition temperatures at 234, 232, 273, 184, 227, and 254 °C, respectively, indicating the drastic substituent effect of adamantyl groups.

1. Introduction.

It is well known that the thermal property is very important to design the novel functional polymeric materials. The incorporations of bulky and rigid alicyclic substituents in the main chain or in the side chain are usually very effective to increase the glass transition temperature (T_g) of various parent polymers [1,2]. In particular, introduction of a bulky adamantane ring gives the high T_gs and excellent thermal stabilities to the parent polymers. In fact, a number of polymers containing adamantyl skeletons in the main chains and in the side chains have been synthesized via the stepwise [3-7], and chain polymerizations [8-15]. As an example of main chain type polymer, we have synthesized a poly(1,3-adamantane) via the ring-opening polymerization of highly strained [3.3.1]propellane derivatives, 1,3-dehydroadamantanes [16]. Furthermore, 1,3-dehydroadamantane undergoes the spontaneous copolymerization with electron-deficient monomers such as acrylonitrile and methyl acrylate in the absence of initiator to form the alternating copolymers containing adamantyl skeleton in the main chain [17].

We have succeeded in the living anionic polymerizations of alkyl methacrylates, 1-adamantyl methacrylates and 3-methacryloyloxy-1,1'-biadamantane, to afford the well-defined polymers exhibiting markedly high T_g values [18]. The resulting polymers possessed the predicted molecular weights based on the molar ratios between monomers and initiators and the narrow molecular weight distributions. In addition, they exhibited remarkably high T_g values and enhanced thermal stabilities derived from the adamantyl groups in each monomer unit. Very recently, we have successfully shown that the anionic polymerization of 2-(1-adamantyl)-1,3-butadiene quantitatively proceeds to afford the well-defined homopolymers and block copolymers carrying the adamantyl group in the side chain [19]. It is well known that styrenes and α-methylstyrenes are also the representative monomers capable of the living anionic polymerization as well as methacrylates and 1,3-dienes.

In this study, we newly designed and anionically polymerized six *para*-substituted styrene derivatives, 4-(1-adamantyl)styrene (**1**) [20], 3-(4-vinylphenyl)-1,1'-biadamantane (**2**) [20], 4-(1-adamantyl)-α-methylstyrene (**3**), 4-(1-adamantyloxy)styrene (**4**), 1-adamantyl 4-vinylbenzoate (**5**), and *N*-(1-adamantyl)-*N*-(4-vinylbenzylidene)amine (**6**), in order to prove the substituent effects of rigid and bulky adamantyl groups on thermal properties of the resulting polystyrenes. In addition, since the monomers possessed the electron-donating group, **4**, or the electron-withdrawing groups, **5** and **6**, the effect of linkage between adamantyl skeleton and the polymer framework on the polymerizability can be clarified as well as the substituent effect on the thermal property.

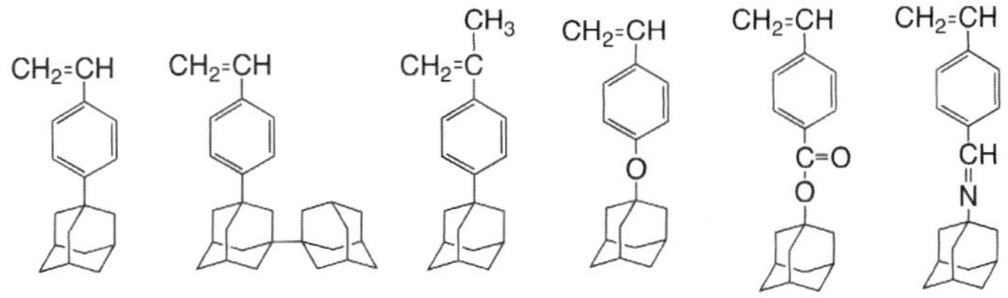

Chart 1. Monomers Containing Adamantyl Groups.

2. Results and Discussion.

2.1. Synthesis Monomers.

Novel monomers, **1** and **3**, were synthesized from 1-adamantyl bromide as illustrated in Scheme 1. In the case of **2**, 3-bromo-1,1'-biadamantane was the starting compound. In these cases, the bulky adamantyl groups were directly connected to the styrene or α-methylstyrene framework. The phenyl ring was introduced on the adamantane ring by the coupling reaction of 1-adamantyl bromide and phenylmagnesium bromide in methylene chloride.

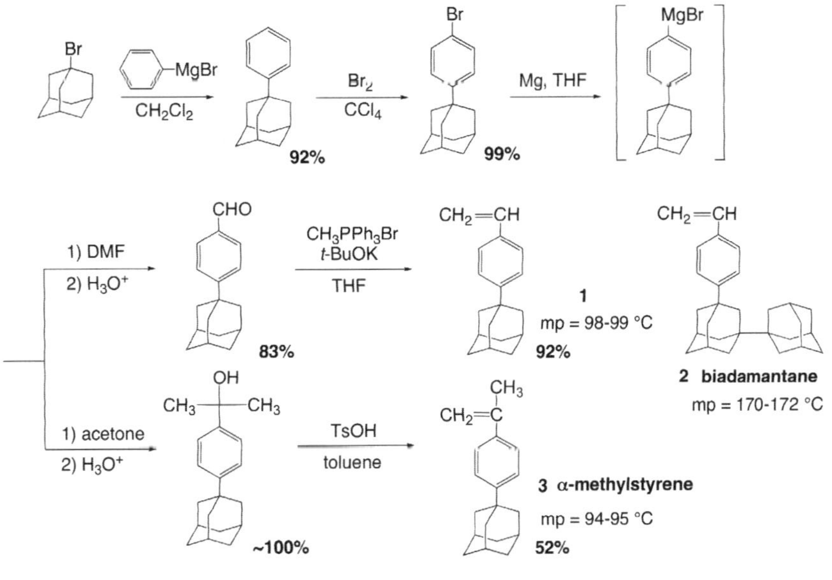

Scheme 1. Synthesis of **1-3**.

Ether type monomer **4** was synthesized as shown in Scheme 2. The adamantyl ether was prepared by the coupling reaction of 1-adamantyl bromide and a phenol derivative, 4-hydroxybenzaldehyde. The resulting benzaldehyde was converted into the corresponding styrene via the Wittig reaction.

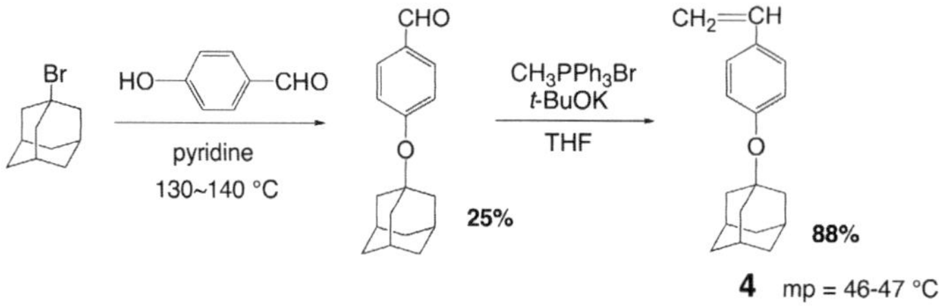

Scheme 2. Synthesis of **4**.

Scheme 3 shows the synthetic routes of ester **5** and imine **6**. In these cases, the Grignard reagent prepared from 4-chlorostyrene was converted into either 4-vinylbenzoic acid or 4-vinylbenzaldehyde. The ester **5** was prepared by the reaction of acid chloride and lithium salt of 1-adamantanol. The condensation between 4-formylstyrene and 1-aminoadamantane gave the imine monomer possessing the C=N linkage.

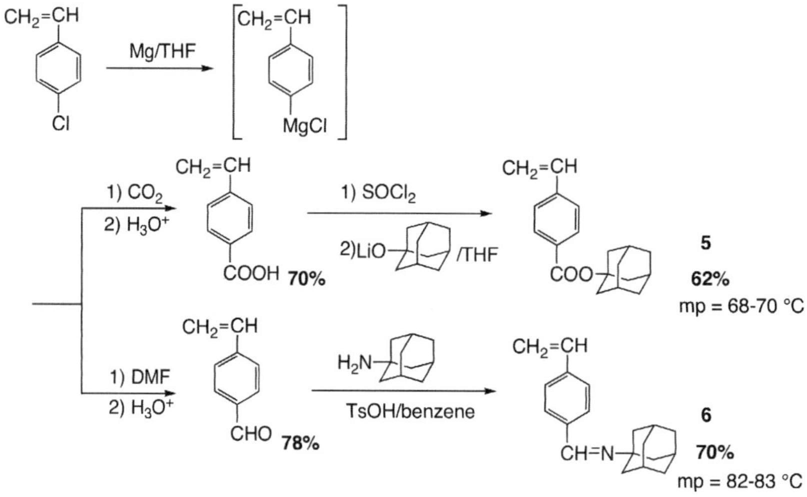

Scheme 3. Synthesis of **5** and **6**.

2.2. Anionic Polymerization.

At first, the anionic polymerization of **1** was carried out in THF at –78 °C as shown in Table 1. The initiators included *sec*-butyllithium (*s*-BuLi), cumylpotassium, potassium naphthalenide, and oligo(α-methylstyryl)dipotassium [20]. During the course of the polymerization, the polymerization system showed a characteristic red color. The coloration disappeared instantaneously by quenching with a small amount of degassed methanol. The complete consumption of monomer was always achieved within 8 h regardless of the counterion of the initiator, and the yield of polymer was quantitative in each case after the termination. The ^{1}H and ^{13}C NMR analysis revealed that the vinyl polymerization of **1** proceeded smoothly. In all cases, SEC curves of the resulting poly(**1**)s were unimodal, and the M_w/M_n values were always within 1.13, indicating the narrow molecular weight distributions. Furthermore, reasonable agreements were observed between the M_n values obtained by RALLS-SEC and the predicted ones based on the feed molar ratios between **1** and initiator. The stability of the propagating carbanion derived from **1** was confirmed by the quantitative initiation efficiency in the postpolymerization. These results clearly show that the polymerization of **1** quantitatively proceeds in a controlled fashion to afford a polymer with well-defined chain structure under the condition employed here. The monomer possessing the bulky biadamantane side chain **2** and α-methylstyrene derivative **3** also underwent the anionic polymerizations to give the tailored polymers quantitatively in THF at –78 °C as shown in Table 1.

Next, we have examined the polymerization of adamantyl ether **4**. The polymerization of **4** occurred with either *s*-BuLi or potassium naphthalenide in THF at –78 °C without difficulty, and the resulting poly(**4**) possessed tailored M_n and narrow molecular weight distribution. On the other hand, no polymeric product of **4** was produced with diphenylmethyllithium in THF at –78 °C after 2 h. The bulky and low nucleophilic carbanion, diphenylmethyllithium, could not initiate the polymerization of **4** at all, since the electron density of vinyl group in the *para*-substituted styrene was increased by the electron-donating property of the adamantyloxy group. This means the low anionic polymerizability of **4** compared with that of styrene due to the substituent effect. By contrast, even the low nucleophilic diphenylmethyl anions can initiate the anionic polymerization of adamantyl ester **5** and adamantyl imine **6**. In these cases, the anionic polymerizability of **5** and **6** was higher than that of styrene, since the electron density of the vinyl groups was significantly reduced by the electron-withdrawing nature of COOAd and CH=N-Ad groups.

Table 1. Anionic Polymerization of **1-6** in THF.

monomer	initiator	temp. °C	time h	yield %	M_n x 10^{-3} calcd	M_n x 10^{-3} obsd[a]	M_w/M_n
1	*s*-BuLi	–78	5 min	100	11	16	1.04
1	*s*-BuLi	–78	1	100	26	26	1.03
1	*s*-BuLi	–78	8	100	54	57	1.03
1	Cumyl-K	–78	3	100	8.4	11	1.05
1	K-Naph	–78	3	100	10	14	1.10
1	K-Naph/α-MeSt	–78	3	100	12	14	1.13
2	*s*-BuLi	–78	16	100	13	11	1.07
2	K-Naph	–78	16	100	6.7	7.2	1.09
3	*s*-BuLi	–78	4	100	11	13	1.05
3	K-Naph	–78	6	100	7.1	7.1	1.08
4	*s*-BuLi	–78	2	100	15	20	1.02
4	Ph_2CHLi	–78	2	0	-	-	-
4	K-Naph	–78	2	100	9.0	10	1.09
5	Ph_2CHLi	–78	4	40	3.7	24[b]	1.20
5	Ph_2CHK	-95	4	100	9.3	14[b]	1.08
6	*s*-BuLi/α-MeSt	–78	24	100	16	32	1.01
6	Ph_2CHK	–78	1	100	20	23	1.02
6	K-Naph/α-MeSt	–78	1	100	29	25	1.03

[a] By RALLS-SEC.

[b] By SEC using polystyrene standards.

The living nature of the polymerization systems of **1-6** allows us to synthesize the block copolymers via the sequential copolymerization with various comonomers such as styrene, α-methylstyrene, and isoprene. For example, a novel ABA triblock copolymer of **1** (A monomer) and isoprene (B monomer) was anionically obtained with *s*-BuLi in cyclohexane at 50 °C via the three stage copolymerization as shown in Scheme 4. The 1,4-polyisoprene segment in the triblock copolymer was quantitatively hydrogenated with *para*-toluenesulfonhydrazide in xylenes at 120 °C to give a triblock copolymer, poly(**1**)-*b*-poly(ethylene-*alt*-propylene)-*b*-poly(**1**). These well-defined

ABA triblock copolymers before and after hydrogenation possessed outer hard A segments and soft middle B segment and showed T_gs at –60 and 220 °C. The higher T_g was corresponding to the poly(**1**) terminal segments as shown later. The analysis of mechanical property also indicates the potential of the block copolymer as a novel thermoplasitic elastomer showing very high service temperature [21].

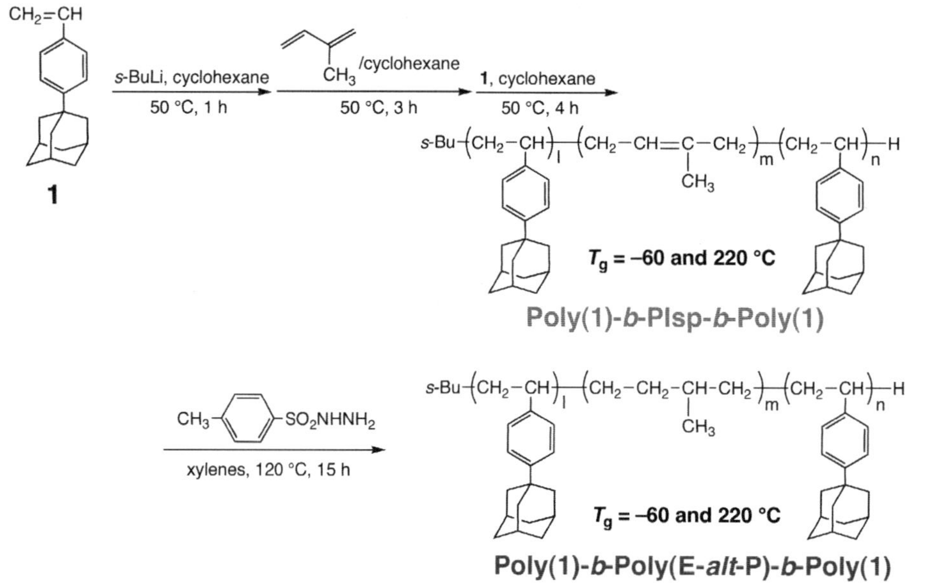

Scheme 4. Synthesis of New Thermoplastic Elastomers.

2.3. Thermal Properties of Polymers.

Thermal stability of the resulting polymers was first investigated by thermogravimetric analysis (TGA) under nitrogen. The polystyrenes, poly(**1**), poly(**2**), poly(**4**), poly(**5**), and poly(**6**), were stable until 340 °C, and poly(**3**), poly(α-methylstyrene) derivative, started to decompose around 290 °C on heating. The initial decomposition temperatures of polymers carrying adamantyl groups were slightly higher than the corresponding non-substituted polymers such as polystyrene and poly(α-methylstyrene).

Next, the T_g values of a series of polymers were analyzed by differential scanning calorimetry (DSC). The T_gs of poly(**1**), poly(**2**), and poly(**3**) were observed at 234, 232, and 274 °C, respectively. These are 110-130 °C higher than the T_g value of parent polystyrene (T_g~100 °C) or poly(α-methylstyrene) (T_g~165 °C). In each case, no melting behavior was observed before the thermal degradation. The T_g value of poly(α-methylstyrene) derivative, poly(**3**), was apparently higher than that of poly(**1**), the polystyrene counterpart. On the other hand, the number of adamantyl group did not affect the T_g value by comparing poly(**1**) and poly(**2**). Figure 1 shows a relationship between T_g and molecular weight of poly(**1**), poly(**4**), poly(**5**), and poly(**6**). Apparently, the T_g values increased with the molecular weight in each case and became almost constant at the high molecular weight region similar to the case of polystyrene. The T_g values of high molecular weight samples for poly(**1**), poly(**4**), poly(**5**), and poly(**6**) reached 234, 184, 227, and 254 °C, respectively. The linkages between adamantyl skeleton and the polystyrene backbone played a very important role to determine the T_gs of polymers as well as the main chain structure.

3. Conclusion.

Herein, the novel polymers having predicted M_ns and narrow molecular weight distributions are synthesized by the living anionic polymerization of a series of novel styrene monomers possessing the adamantyl skeletons. The incorporation of bulky and stiff adamantyl and 1,1'-biadamantyl moieties unequivocally induces the decreasing of chain mobility to result in the drastic T_g enhancement of a series of novel 4-substituted polystyrenes.

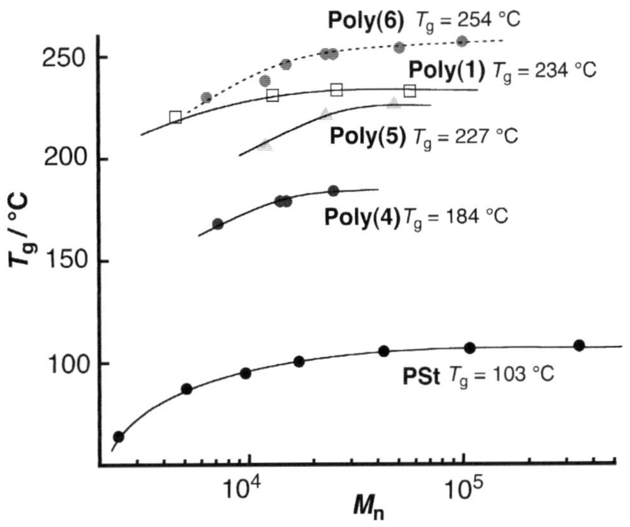

Figure 1. Relationship between T_g and M_n of Polymers.

[1] Cypcar C, Camelio P, Lazzeri V, Mathias L J 1996 *Macromolecules* **29** 8954
[2] Yu J M, Dubois Ph, Jérôme R 1996 *Macromolecules* **29** 7316
[3] Jensen J J, Grimsley M, Mathias L J 1996 *J. Polym. Sci. Part A* **34** 397
[4] Mathias L J, Lewis C M, Wiegel K N 1997 *Macromolecules* **30** 5970
[5] Fukukawa K, Shibasaki Y, Ueda M 2004 *Macromolecules* **37** 8256
[6] Seino H, Mochizuki A, Ueda M 1999 *J. Polym. Sci., Part A. Polym. Chem.* **37** 3584
[7] Chern Y T and Shiue H C 1997 *Macromolecules* **30** 4646
[8] Otsu T, Matsumoto A, Horie A, Tanaka S 1991 *Chem. Lett.* 1145
[9] Matsumoto A, Tanaka S, Otsu T 1991 *Macromolecules* **24** 4017
[10] Feng F, Mitsuishi M, Miyashita T 1999 *Langmuir* **15** 8673
[11] Teraguchi M and Masuda T 1999 *J. Polym. Sci., Part A. Polym. Chem.* **37** 4546
[12] Acar H Y, Jensen J J, Thigpen K, McGowen J A Mathias L J 2000 *Macromolecules* **33** 3855
[13] van Reenen A J, Mathias L J, Coetzee L 2004 *Polymer* **45** 799
[14] Hashimoto T, Makino Y, Urushisaki M, Sakaguchi T 2008 *J. Polym. Sci., Part A. Polym. Chem.* **46** 1629
[15] Kavitha A A and Singha N K 2008 *J. Polym. Sci., Part A. Polym. Chem.* **46** 7101
[16] Ishizone T, Matsuoka S, Sakai S, Harada W, Tajima H 2004 *Macromolecules* **37** 7069
[17] Matsuoka S, Ogiwara N, Ishizone T 2006 *J. Am. Chem. Soc.* **128** 8708
[18] Ishizone T, Tajima H, Torimae H, Nakahama S 2002 *Macromol. Chem. Phys.* **202** 2375
[19] Kobayashi S, Kataoka H, Ishizone T, Kato T, Ono T, Kobukata S, Arimoto K, Ogi H *Reactive and Functional Polym.* in press.
[20] Kobayashi S, Matsuzawa T, Matsuoka S, Tajima H, Ishizone T 2006 *Macromolecules* **39** 5979
[21] Kobayashi S, Kataoka H, Ishizone T, Kato T, Ono T, Kobukata S, Ogi H 2008 *Macromolecules* **41** 5502

Journal of Physics: Conference Series **184** (2009) 012018 doi:10.1088/1742-6596/184/1/012018

Static and Dynamic Light Scattering Studies on Dilute Polyrotaxane Solutions

Tetsuya Kume[1], Jun Araki[2*], Yasuhiro Sakai[1], Koichi Mayumi[1], Masatoshi Kidowaki[1], Hideaki Yokoyama[1] and Kohzo Ito[1*]

[1] Department of Advanced Materials Science, Graduate School of Frontier Sciences, University of Tokyo, 5-1-5 Kashiwanoha, Kashiwa City, Chiba 277-8561, Japan

[2] International Young Researchers Empowerment Center, Shinshu University, Tokida 3-15-1, Ueda City, Nagano 384-8567, Japan

E-mail: jun@shinshu-u.ac.jp (Jun Araki), kohzo@molle.k.u-tokyo.ac.jp (Kohzo Ito)

Abstract. Static and dynamic light scattering measurements were performed for dilute polyrotaxane solutions in different types of solvent systems, i.e. dimethylacetamide (DMAc) or dimethylformamide (DMF) containing 1–6 wt% lithium chloride (LiCl), 1 M aqueous sodium hydroxide (NaOH) and dimethylsulfoxide (DMSO). No aggregation of the polyrotaxane in DMF/LiCl was confirmed in the present study. Radius of gyration of the dissolved polyrotaxane was largest in NaOHaq., followed by values in amide solvents/LiCl and that in DMSO, and was probably dominated not by Coulombic repulsion but by the mutual affinity between solvent and polyrotaxane. Ratio of radius of gyration to hydrodynamic radius suggested the flexible random-coiled conformation in DMSO and relatively more extended, semi-flexible ones in amide solvents/LiCl and NaOHaq. The obtained values of second virial coefficient and weight average molecular weight seemed to be affected by a potential change in differential refractive index increments, caused by selective macrocationization or ionization.

1. Introduction

Polyrotaxane is one of the most vigorously investigated supramolecules in which many cyclic molecules are penetrated with a linear molecule [1–4]. One of the typical polyrotaxane molecules consists of α-cyclodextrins (CDs) as cyclic molecules and poly(ethylene glycol) (PEG) as a linear molecule [5,6]. Recent attentions have been focused on applications of the PEG/CD polyrotaxane to construct a wide variety of functionalized materials, such as a nano-sized tubular molecule [7, 8], molecular machines [9], an advanced drug-delivery system [10] and preparation of topologically crosslinked polyrotaxane gels [11–13]. The polyrotaxane gels called as "slide-ring gels" or "sliding gels" can achieve an extremely high degrees of elongation or swelling due to free sliding movements of the internal crosslinked cyclodextrins [11–14].

Physical and chemical properties of these polyrotaxane-based functional materials, which are prepared *via* solution state, can be affected by the state of the starting solutions. The physical properties of polyrotaxane solutions are, therefore, of great interest. Previous studies suggested a wide

[*] To whom any correspondence should be addressed.

variety of intriguing phenomena of polyrotaxane solutions; namely, solvent-dependent changes in aggregation of CDs [15, 16], drastic difference between mechanical properties of slide-ring gels containing different solvents [13, 17], a rigid rod-like structure at a high degree of inclusion and its change into flexible structure by chemical modifications [18, 19], and a surprisingly strong thixotropy of a concentrated polyrotaxane solution in DMSO [20]. In spite of these energetic studies, surprisingly less efforts have been focused on the difference in molecular conformation of a single polyrotaxne chain in a dilute solution with several types of solvents. Such investigations can be now examined thanks to the recent discovery of a variety of good solvents for polyrotaxane [17, 21–23]. Availableness of new solvents stimulated us to investigate solvent-molecule interaction in several dilute polyrotaxane solutions.

Static light scattering (SLS) and dynamic light scattering (DLS) techniques are powerful tools for obtaining information on conformation of polymer chains in a dilute solution [24]. These measurements provide polymer parameters such as radius of gyration R_g, second virial coefficient A_2, weight average molecular weight M_w, diffusion coefficient D and hydrodynamic radius R_h. Since there has been unfortunately no SLS and DLS study on the dilute polyrotaxane solutions, we examined an evaluation of the molecular parameters of the PEG/CD polyrotaxane in different solvent systems.

2. Materials and Methods

Reagents other than polyrotaxane were purchased from Wako Pure Chemical Industries, Ltd. All reagents were of experimental grade and used without further purification. Polyrotaxane was prepared according to the previous method [25]. The inclusion ratio, i.e. percentage of the PEG axis covered with CDs, was calculated as 26% from ^{1}H-NMR measurement. From gel permeation chromatography measurements under the conditions described before [20], weight average molecular weight and polydispersity of the synthesized polyrotaxane was calculated as 1.40×10^5 and 1.29, respectively.

As solvents for light scattering measurements, the following solvent systems were selected: dimethylsulfoxide (DMSO), 1 M aqueous solution of sodium hydroxide (1 M NaOHaq.), dimethylacetamide (DMAc) containing lithium chloride (LiCl) and dimethylformamide (DMF) containing LiCl. Prior to the light scattering measurements, qualitative solubility of the polyrotaxane in these solvent systems was visually examined after mixing ca. 20 mg of the polyrotaxane with 1 mL of the above solvents, adjusting LiCl content at 0.5–8 wt% in the case of amide solvent/LiCl systems.

SLS and DLS measurements were performed with ALV–5000E (ALV–GmbH, Langen, Germany), using He–Ne laser with wavelength λ = 632.8 nm, at 25 °C and at the measurement angles between 30° and 150°. Rayleigh ratio was calibrated with the value for toluene (1.3522×10^{-5} cm^{-1}). The sample solutions at 0.3–5.0 mg/ml polyrotaxane content were subjected to measurements after optical clarification with a 0.20 μm PTFE filter. Differential refractive indices dn/dc of the solutions were measured with a differential refractometer (DRM–1021, Otsuka electronics Co. Ltd.). Viscosity of the solvent was measured at 25 °C by an Ubbelohde viscometer.

The data obtained by SLS measurements were analyzed with Zimm plots according to the following eqs 1–3;

$$\left.\frac{Kc}{R_\theta}\right|_{c\to 0} = \frac{1}{M_w}\left(1+\frac{1}{3}R_g^2q^2+\cdots\right) \quad (1)$$

$$\left.\frac{Kc}{R_\theta}\right|_{\theta\to 0} = \frac{1}{M_w}+2A_2c+\cdots \quad (2)$$

$$K = \frac{4\pi^2n^2}{N_A\lambda^4}\left(\frac{dn}{dc}\right)^2 \quad (3)$$

where K is an optical constant, c polymer concentration, n a refractive index of the measured solvent, $q = (4\pi n/\lambda)\sin(\theta/2)$ the magnitude of a scattering vector, λ a wavelength of the incident light under vacuum, θ a measurement angle and *dn*/*dc* a differential refractive index of the solution. These analyses give the values of M_w, R_g and A_2.

Calculation of the initial relaxation speed Γ obtained with eq 4 gives diffusion coefficient D of the polymer, which was translated into hydrodynamic radius R_h according to eq 5;

$$\Gamma = Dq^2 \qquad (4)$$

$$R_h = (kT / 6\pi\eta D)(1+k_D c) \qquad (5)$$

where k_D represents the effect of the interaction, k is the Boltzmann constant and η solvent viscosity.

3. Results and Discussion

3.1. Dissolution Mechanism of Polyrotaxane in Various Solvent Systems

In the present study, we found an excellent dissolution of the PEG/CD polyrotaxane in an unreported type of solvent, i.e. DMF containing LiCl. Qualitative dissolution tests with varied amount of LiCl revealed that the DMF containing LiCl above 1 wt% and DMAc containing LiCl above 0.5 wt% readily dissolve the polyrotaxane, and the dissolution in the former was considerably rapid (typically within 15 minutes at room temperature) compared to that in the latter. The solvents with lower salt content did not give complete dissolution.

Based on the previous investigation [26] on dissolution of cellulose, i.e. a highly hydrogen-bonded crystalline polysaccharide, as well as the examinations by Nakamura et al [27], the dissolution mechanism of the polyrotaxane in DMF/LiCl and DMAc/LiCl can be explained by complexation of Li^+ and solvent molecules (DMAc or DMF), followed by selective binding of the formed macrocation to hydroxyl groups of the polyrotaxane and resulting interruption of intermolecular hydrogen bonding [4, 21, 22]. The polyrotaxane molecules dissolved in these solvent systems are considered to be cationically charged due to this macrocation complexation. Dissolution of the polyrotaxane in 1 M NaOHaq. is also dominated by ionization of the polyrotaxane molecules *via* dissociation of hydroxyl groups above pH 12, the pK_a of the hydroxyl groups in CD [5, 6]. Considering this mechanism, the polyrotaxane molecules in 1 M NaOHaq. are anionically charged. In contrast to the dissolution in the above solvent systems, the polyrotaxane dissolved in DMSO is considered to be nonionic (unionized). Analogous to the dissolution mechanism of CDs in DMSO, reported by Shikata et al. [28], the polyrotaxane is dissolved by solvation of DMSO molecules hindering hydrogen bond formation.

3.2. Light Scattering Measurements

Figure 1 shows a Zimm plot of the polyrotaxane solution in DMF/LiCl (2 wt%), in which Kc/R_θ was plotted against $\sin^2(\theta/2) + 0.2c$. According to eq 1, extrapolation of Kc/R_θ to $c \rightarrow 0$ and $\theta \rightarrow 0$ gives $1/M_w$, while the initial slope at $\theta \rightarrow 0$ reflects A_2. The value of R_g is calculated from the slope of $\sin^2(\theta/2)$ at $c \rightarrow 0$.

Figure 2 shows a plot of diffusion coefficient D, measured in the same solvent by DLS and plotted against polyrotaxane concentration c. Each D value at every c was calculated according to eq. 4, considering dependence of Γ on q (and therefore on θ) by measuring Γ at various q. According to eq. 5, extrapolation to $c \rightarrow 0$ gives a value of D corresponding to hydrodynamic radius R_h of the dissolved polyrotaxaene. These measurements and calculations were also performed for the other solvent systems, i.e. DMF/LiCl (1, 4 and 6wt%), DMAc/LiCl (2wt%), 1 M aqueous NaOH solution and DMSO. The obtained parameters were summarized in Table 1, together with values of refractive index increment *dn*/*dc* and solvent viscosity η.

The values for R_g exhibited a wide variation with the types of solvents. The R_g measured in 1 M NaOHaq. showed the largest value, followed by those in DMF/LiCl and DMAc/LiCl, and the smallest value was observed in the measurement in DMSO. The value in DMSO, 16.9 nm, is comparable to that obtained by the measurement in DMSO after the hydroxypropylation of the same polyrotaxane (19.5 ± 1 nm) [29], and also well correspond to that reported by Fleury et al. (14.5 ± 10 nm) [19] for the PEG/CD polyrotaxane with similar degree of inclusion (number of CD is 75, correspond to inclusion ratio of 33%).†

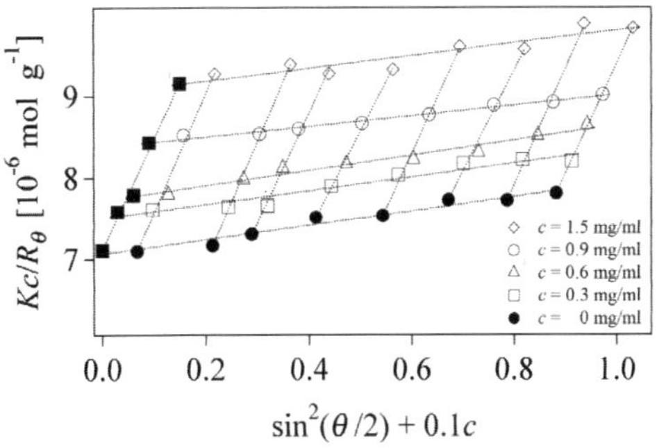

Figure 1. Typical Zimm plot for the polyrotaxane solution in DMF/LiCl (2wt%). (filled circles) extrapolations to zero concentration, (filled squares) extrapolations to zero angle.

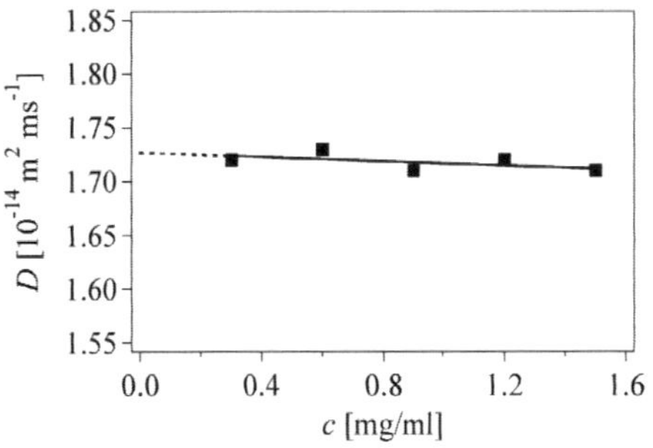

Figure 2. Diffusion coefficients D obtained by DLS measurements of polyrotaxane solution in DMF/LiCl (2wt%) and plotted against polymer concentration c.

Table.1 Viscosity of the solvent η, refractive index increment dn/dc, radius of gyration R_g, hydrodynamic radius R_h, $\rho = R_g/R_h$, weight average molecular weight of the polyrotaxane M_w and second virial coefficients A_2 determined in various solvent systems.

Solvent	η, mPa•s	dn/dc, ml/g	R_g, nm	R_h, nm	κ^{-1}, nm	ρ	A_2, mol·cm^3/g^2	M_w, g/mol
DMAc/LiCl (2wt%)	1.73	9.59×10^{-2}	19.4	10.6	0.31	1.83	5.41×10^{-4}	1.75×10^5
DMF/LiCl (1wt%)	1.04	1.04×10^{-1}	19.1	12.2	0.43	1.57	1.63×10^{-3}	1.07×10^5
DMF/LiCl (2wt%)	1.52	1.07×10^{-1}	21.1	8.3	0.31	2.54	6.90×10^{-4}	1.42×10^5
DMF/LiCl (4wt%)	2.01	1.10×10^{-1}	20.9	9.3	0.22	2.25	1.39×10^{-4}	1.08×10^5
DMF/LiCl (6wt%)	2.58	1.13×10^{-1}	25.6	10.7	0.18	2.39	2.09×10^{-3}	9.0×10^4
1 M NaOHaq.	1.00	1.43×10^{-1}	29.8	13.9	0.22	2.14	4.67×10^{-4}	1.86×10^5
DMSO	2.00	5.84×10^{-2}	16.9	10.4	-	1.62	5.28×10^{-4}	1.12×10^5

As stated in the above section, polyrotaxane molecules dissolved in NaOHaq., DMF/LiCl and DMAc/LiCl are ionized. Such ionization mechanisms are readily expected to affect net R_g values of the dissolved polyrotaxanes through short- and long-range interactions. To our surprise, the recent results by Mayumi et al. [30] evidently contradict the above forecast; namely, SANS measurement of the nonionic water-soluble polyrotaxane (hydroxypropylated polyrotaxane) in D_2O and 1 M NaOD gave almost identical scattering profiles, in spite of the ionization of the polyrotaxane in the latter solvent [30]. Their results clearly indicate the identical R_g in both solvents, and subsequently no effect of the electrical ionization on both short- and long-range interactions. Considering the results by Mayumi et al. together with ours, the present variations in R_g with change in the type of solvents seems to be affected mainly by the mutual affinity between solvent and polyrotaxane, which should be evaluated by the parameter A_2.

† Change in R_g value with the difference in the length of PEG axis should be considered. R_g is proportional to (degree of polymerization)$^{3/5}$ and therefore to (chain length)$^{3/5}$ in a good solvent. The contour length of polyrotaxane is equal to that of PEG axis. Values of number average molecular weight M_n of PEG used by Fleury et al. and us are 20000 [19] and 28600 [25], respectively. R_g value of 14.5 nm for the polyrotaxane with PEG20000 axis corresponds to that of 18.0 nm for the polyrotaxane with PEG28600 axis, which well agrees with our experimental results (16.9 nm).

Our measurements on polyrotaxane solution in DMF/LiCl series also showed almost no change in R_g values with the amount of added LiCl. This behavior is reasonably explained by evaluating Debye length κ^{-1}, i.e. a distance which Coulombic repulsive force is shielded, by the following eqs [31];

$$\kappa^2 = 8\pi d_B C_S \qquad (7)$$

$$d_B = \frac{e^2}{4\pi\varepsilon\varepsilon_0 kT} \qquad (8)$$

where d_B is Bjerrum length (ca. 0.7 nm in the case of aqueous systems), C_S a salt concentration, e the elementary charge, T the absolute temperature, ε the relative dielectric constant of solution, and ε_0 the vacuum permittivity. The values of κ^{-1} in the used solvent systems except for DMSO are also summarized in Table 1. All of the obtained values are less than 1 nm, indicating almost shielded Coulombic interactions in these solvents. This may be the reason for the above-mentioned independence of R_g values on ionization [30], especially for no change in short-range interactions.

Hydrodynamic radii R_h of polyrotaxane obtained in DMAc/LiCl and DMF/LiCl seemed to be relatively lower than those in NaOHaq. and DMSO, but slightly increased with the increase in LiCl content, probably due to enhanced hydrodynamic interactions by larger degree of macrocationization. Another study on DLS measurements of cellulose dissolved in aqueous alkali metal complex solvents also suggests the quite complicated effect of selective ion/polymer binding on hydrodynamic interactions [32].

Ratio of R_g to R_h, $\rho = R_g/R_h$, is a parameter which provides a significant information on the conformation of a polymer molecule in dilute solution. While suspension of spherulitic particles theoretically gives $\rho = 0.775$ [33], a random-coiled polymer in a theta solvent and a good solvent shows the ρ values of 1.506 and 1.862, respectively [34], and the higher values are observed for semi-flexible polymer solutions. The ρ values calculated for all solvent system used are listed in Table 1. The ρ values calculated for NaOHaq. and DMF/LiCl with LiCl content over 2wt% are all beyond 2, and indicate the semi-flexible and relatively extended chain conformation of polyrotaxane molecules in these solvent systems. The ρ values for DMSO, DMAc/2wt% LiCl and DMF/1wt% LiCl (between 1.5 and 1.9), which are comparable to that of the random-coils in a good solvent (1.862), suggest that the polyrotaxane behaves as a flexible chain in these solvents.

In contrast to the estimation of R_g and R_h that were independent on *dn/dc*, Some considerations should be taken for the interpretation of M_w and A_2, since these values can be affected by several difficulties such as a fluctuation of *dn/dc* values with types of solvents [35] or selective and localized macrocation complexation [32], especially in the case of electrolyte solvents. Therefore, only a qualitative evaluation for the tendency of M_w and A_2 are discussed in the following text.

The values of A_2 for DMF/LiCl series showed a slight increase at high amount of added LiCl. This tendency implies the increase in solvent-polymer compatibility with increase in the degree of macrocation complexation, but contradicts with above-mentioned no change in R_g values, which is considered to be affected mainly by A_2 in our speculation.

The values of M_w calculated from SLS results exhibited a wide variation with the types of solvents used, despite the use of the same polyrotaxane sample for all measurements. One of the reasons is related to the above-mentioned selective complexation or ionization of polyrotaxane, resulting in difference between refractive indices of the solution close to and apart from polymer. It can be concluded that the SLS measurements in DMSO give the most reliable M_w value, considering no effect of selective cation binding and a ρ value indicating flexible conformation. The M_w of the polyrotaxane calculated from SLS measurement of a DMSO solution (1.12×10^5) actually shows a value very similar to that calculated from GPC measurements (1.40×10^5, show the experimental section), as well as that of the similar previous sample obtained by ^{1}H NMR and GPC (1.18×10^5) [25].

ACKNOWLEDGMENT

This work was partially supported by the Ministry of Education, Science, Sports and Culture, Japan (Grant-in-Aid for Scientific Research (S), 2008-2012, No. 20221005), and Grant-in-Aid for Scientific Research on Priority Area “Soft Matter Physics”. Part of this study was performed through Special Coordination Funds for Promoting Science and Technology of the Ministry of Education, Culture, Sports, Science and Technology, the Japanese Government.

References

[1] Harada A, Hashidzume A and Takashima Y 2006 *Adv. Polym. Sci.* **201** 1–43
[2] Takata T, Kihara N and Furusho Y 2004 *Adv. Polym. Sci.* **171** 1–76
[3] Ito K 2007 *Polym. J.* **39** 489–499
[4] Araki J and Ito K 2007 *Soft Matter* **3** 1456–1473
[5] Harada A, Li J and Kamachi M 1992 *Nature* **356** 325–327
[6] Harada A, Li J, Nakamitsu T and Kamachi M 1993 *J. Org. Chem.* **58** 7524–7528
[7] Harada A, Li J and Kamachi M 1993 *Nature* **364** 516–518
[8] Samitsu S, Araki J, Shimomura T and Ito K 2008 *Macromolecules* **41** 5385–5392
[9] Schalley C A, Beizai K and Vögtle F 2001 *Acc. Chem. Res.* **34** 465–476
[10] Ooya T and Yui N 1998 *Macromol. Chem. Phys.* **199** 2311–2320
[11] Okumura Y and Ito K 2001 *Adv. Mater.* **13** 485–487
[12] Fleury G, Schlatter G, Brochon C and Hadziioannou G 2005 *Polymer* **46** 8494–8501
[13] Fleury G, Schlatter G, Brochon C, Travelet C, Lapp A, Linder P and Hadziioannou G 2007 *Macromolecules* **40** 535–543
[14] Shinohara Y, Kayashima K, Okumura Y, Zhao C, Ito K and Amemiya Y 2006 *Macromolecules* **39** 7386–7391
[15] Karino T, Okumura Y, Zhao C, Ito K and Shibayama M 2005 *Macromolecules* **38** 6161–6167
[16] Karino T, Okumura Y, Ito K and Shibayama M 2004 *Macromolecules* **37** 6177–6182. Karino T, Okumura Y, Ito K and Shibayama M 2005 *Macromolecules* **38** 1035 (Erratum)
[17] Samitsu S, Araki J, Kataoka T and Ito K 2006 *J. Polym. Sci. B Polym. Phys.* **44** 1985–1994
[18] Jarroux N, Guegan P, Cheradame H and Auvray L 2005 *J. Phys. Chem. B* **109** 23816–23822
[19] Fleury G, Brochon C, Schlatter G, Bonnet G, Lapp A and Hadziioannou G 2005 *Soft Matter* **1** 378–385
[20] Araki J and Ito K 2007 *Polymer* **48** 7139–7144
[21] Araki J and Ito K 2006 *J. Polym. Sci. A Polym. Chem.* **44** 532–538
[22] Araki J and Ito K 2006 *J. Polym. Sci. A Polym. Chem.* **44** 6312–6323
[23] Araki J, Kataoka T and Ito K 2007 *J. Appl. Polym. Sci.* **105** 2265–2270
[24] Burchard W 1983 *Adv. Polym. Sci.* **48** 1–124
[25] Araki J, Zhao C M and Ito K 2005 *Macromolecules* **38** 7524–7527
[26] McCormick C L, Callais P A and Hutchinson B H 1985 *Macromolecule* **18** 2394–2401
[27] Nakamura T 1975 *Bull. Chem. Soc. Jpn.* **48** 1447–1451
[28] Shikata T, Takahashi R, Onji T, Satokawa Y and Harada A 2006 *J. Phys. Chem. B* **110** 18112–18114
[29] Mayumi K, Osaka N, Endo H, Yokoyama H, Sakai Y, Shibayama M and Ito K 2008 *Macromolecules* **41** 6480–6485
[30] Mayumi K, Osaka N, Endo H, Yokoyama H, Shibayama M and Ito, K Manuscript in preparation.
[31] Ookubo N, Hirai Y, Ito K and Hayakawa R 1989 *Macromolecules* **22** 1359–1366
[32] Saalwächter K, Burchard W, Klüfers P, Kettenbach G, Mayer P, Klemm D and Dugarmaa S 2000 *Macromolecules* **33** 4094–4107
[33] Rubinstein M and Colby R H 2003 *Polymer Physics.* Oxford University Press Chapter 8
[34] Akcasu A Z, Benmouna M and Alkhafaji S 1981 *Macromolecules* **14** 147–150
[35] Ball V and Ramsden J J 1998 *Biopolymers* **46** 489–492

Formation of Polystyrene/Poly(methyl methacrylate) Hetero-arm Star-Like Nanogels from Complementarily Reactive Well-Defined Diblock Copolymers

Y Amamoto,[1] H Otsuka,[1,2] and A Takahara[1,2]

[1]Graduate School of Engineering, Kyushu University, 744 Motooka, Nishi-ku, Fukuoka, 819-0385, Japan
[2]Institure for Materials and Chemistry, Kyushu University, 744 Motooka, Nishi-ku, Fukuoka, 819-0385, Japan

E-mail: otsuka@ms.ifoc.kyushu-u.ac.jp

Abstract. Formation of star-like nanogels with two different arms via cross-linking reaction of complementarily reactive diblock copolymers was successfully accomplished. The two types of diblock copolymers, consisting of poly(methyl methacrylate) (PMMA) or polystyrene (PSt) block and alkoxyamine-based cross-linkable block, were prepared by atom transfer radical polymerization (ATRP) methods. The cross-linking reactions were carried out by merely heating their mixture, and traced by gel permeation chromatography (GPC) and multi-angle light scattering (MALS) measurements. The diblock copolymers were reacted in complementarily reactive systems, showing that all star-like nanogels have necessarily two types of arms as PMMA and PSt chains.

1. Introduction

Star polymers have been synthesized by expanded methods using controlled polymerization, and several types of star-like architectures have been reported.[1–5] Designing different kinds of arms in one star polymer is one of the most remarkable challenges, because their physicochemical properties might be different from those of linear polymers. So far, some star polymers have been synthesized by so-called "mixed method" in star polymer synthesis. Hadjichristidis et al. reported 3- or 4-miktoarm (hetero-arm) star polymers by using coupling reactions of chain ends in anionic polymerization and chlorosilane derivatives.[2,3] Higashimura et al. represented hetero-arm amphiphilic star-shaped polymers via cationic polymerization from core-functional star polymers.[4] Furthermore, Hirao et al. synthesized asymmetric star-branched polymers by means of anionic polymerization using 1,1-diphenylethylene derivatives.[5]

The authors have reported the formation of star-like nanogels via cross-linking reaction of diblock copolymers with alkoxyamine units in their side chains.[6] An alkoxyamine is an adduct of styryl radical and nitroxide radical behaving as a typical organic molecule under normal condition, but plural alkoxyamines can dissociate into radicals and exchange their units under heating condition.[7,8] Because of such reactivity, the central C–O bonds in alkoxyamine can be treated as dynamic covalent bonds.[9] In addition, cross-linking reaction occurred by designing and heating linear polymers with alkoxyamine units in their side chains [10], and nanogels with star-like structure were formed in the case of diblock copolymer systems. Interestingly, because an alkoxyamine consists of different types of radical species, the complementarily reactive systems could be constructed by controlling their direction. So far, we have reported the formation of macroscopic gels and star-like nanogels using two types of complementarily reactive linear polymers consisting of poly(methyl methacrylate) (PMMA) derivatives.[10–12] Such systems essentially possess to combine two different kinds of polymers into any organized structure. Herein, a novel methodology of formation of star-like architecture with different arms as PMMA and polystyrene (PSt) is proposed.

2. Experimental

2.1. Measurement
The relative number average molecular weight (M_n) and molecular weight distribution (M_w/M_n) of polymers were estimated by gel permeation chromatographic (GPC) analysis, which was carried out at 40 °C on TOSOH HLC-8220 GPC system equipped with a guard column (TOSOH TSK guard column Super H-L), three columns (TOSOH TSK gel SuperH 6000, 4000, and 2500), and a UV-vis detector (TOSOH UV-8020). The absolute molecular weight of nanogels was measured by gel permeation chromatography – multiangle light scattering (GPC-MALS) in THF at 40 °C on a Dawn EOS instrument (Wyatt Technology; Ga-As laser, λ = 690 nm). The specific refractive index increment ($\partial n/\partial c$) which is necessary for the analysis of GPC-MALS was measured with Otsuka Electric DRM-1020 (40 °C, λ = 633 nm). The measured $\partial n/\partial c$ of diblock copolymers and nanogels in THF was 0.0937-0.125 cm^3 g^{-1} corresponding to mixing ratio.

2.2. Preparation of diblock copolymers
Diblock copolymer **1** was synthesized by random copolymerization of methyl methacrylate (MMA) and methacrylic ester with alkoxyamine units from PSt (M_n = 18 800, M_w/M_n = 1.18) with bromine at the chain end by the ATRP method, as shown in Scheme 1. The polymerization was carried out by using CuCl/4,4'-dinonyl-2,2'-dipyrydyl (dNbpy) as catalyst in anisole at 50 °C for 24 h. After the polymerization, the catalyst was removed by an alumina column, and the crude was purified by reprecipitation in acetonitrile/hexane mixed solvent to obtain polymer as white precipitation (M_n = 27 700, M_w/M_n = 1.37). Diblock copolymer (M_n = 26 700, M_w/M_n = 1.18) **2** was synthesized as previously reported [10] from PMMA-Br (M_n = 21 500, M_w/M_n = 1.18) prepolymer.

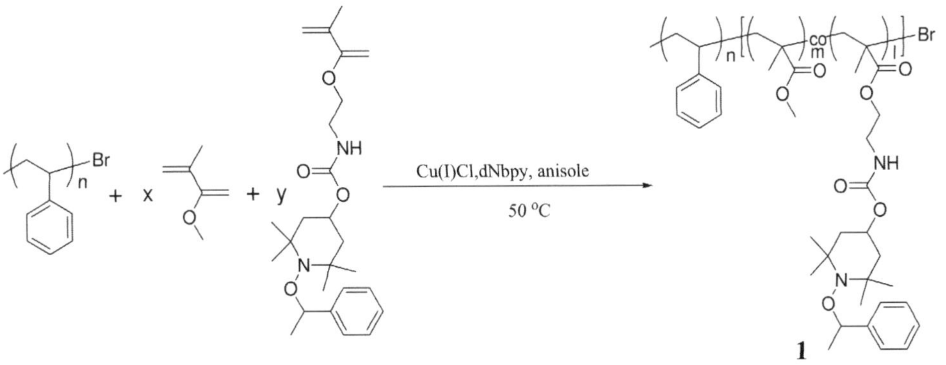

Scheme 1. Preparation of diblock copolymer **1** by random copolymerization of MMA and methacrylic ester with alkoxyamine via the ATRP method.

2.3. Cross-linking Reaction
A 10 wt % anisole solution of mixture of diblock copolymers **1** and **2** was charged into a glass tube, degassed by seven freeze-pump-thaw cycles, and sealed off under vacuum. The solution was heated at 100 °C for 24 h to afford a colorless gel. The gel product was washed with chloroform and dried in vacuo to afford a white solid of cross-linked polymer **3** in quantitative yield.

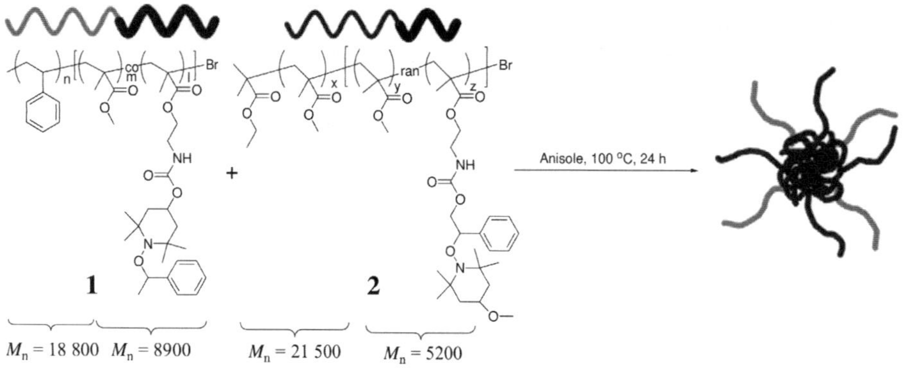

Scheme 2. Formation of hetero-arm star-like nanogels via radical crossover reaction of diblock copolymers **1** and **2** with alkoxyamine units in their side chains.

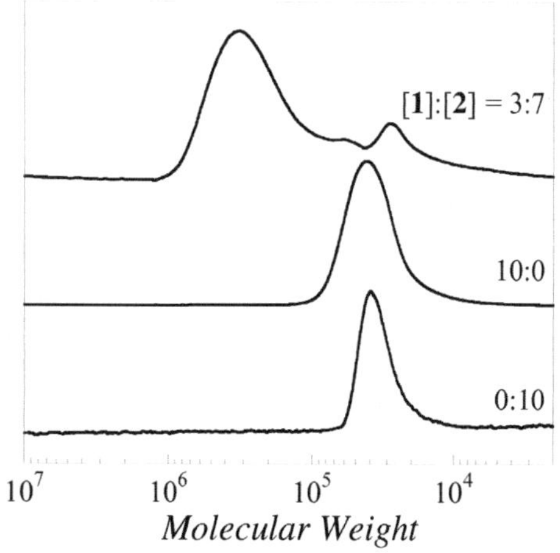

Figure 1. GPC curves of reaction products after heating a mixture of diblock copolymers 1 and 2 ([1]:[2] = 3:7), and each diblock copolymer ([1]:[2] = 10:0, 0:10) in anisole (10 wt %) for 24 h at 100 °C.

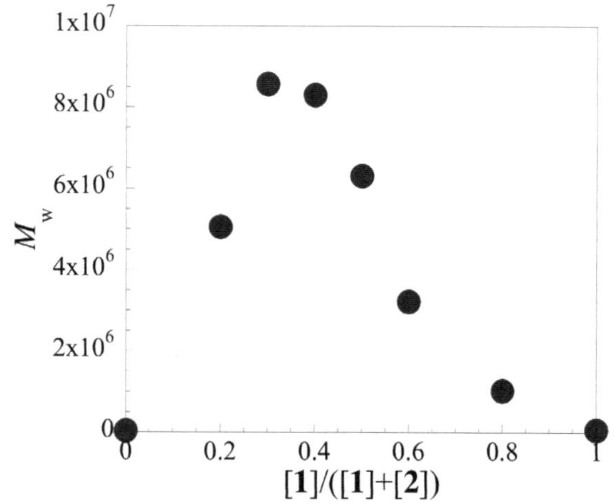

Figure 2. Mixing ratio dependence of M_w of reaction product after heating mixtures of diblock copolymers **1** and **2**, and each diblock copolymer in anisole (10 wt %) for 24 h at 100 °C.

3. Results and discussion

Two types of diblock copolymers **1** and **2** with alkoxyamine units connected at different positions were synthesized by ATRP methods. The reactions were carried out by copolymerization of MMA and methacrylic ester with alkoxyamine units from PSt and PMMA prepolymers, respectively, and the connecting diblock copolymers with relatively low molecular weight distributions were obtained. In the case of diblock copolymer **1**, a halogen exchange method was used,[13] and unreacted PSt homopolymer was removed by reprecipitation using acetonitrile/hexane mixed solvent. The molecular weight distribution of diblock copolymer **1** (M_w/M_n = 1.37), however, was larger than that of **2** (M_w/M_n = 1.18), because the polymerization rate might be larger than initiation reaction rate for the preparation of **1**. Cross-linking reactions were carried out by heating mixtures of diblock copolymers **1** and **2** in anisole at 100 ºC for 24 h to exchange alkoxyamine units in the radical process, as shown in Scheme 2. After heating, no gelation of the system was observed in spite of higher concentration condition such as 10 wt %, because the PSt and PMMA chains without alkoxyamine units depressed the macroscopic gelation. Furthermore, the coloration corresponding to a generation of nitroxide radicals was not also observed, which showed that the exchange reaction of alkoxyamine units proceeded without detectable side reaction.

Figure 1 shows GPC curves after heating the mixture ([**1**]:[**2**] = 3:7) and each diblock copolymer (**1** and **2**) in anisole at 100 °C for 24 h. In the case of heating each diblock copolymer, the GPC curves hardly changed. On the other hand, the higher molecular weight peaks, corresponding to cross-linked polymer, could be obtained without exceeding exclusion limit in the case of mixing conditions. These results indicated that cross-linking reaction occurred by exchange reaction of alkoxyamine units, and the excessive gelation was depressed by existence of PMMA and PSt chains.

In order to investigate the complementarily reactive system, the cross-linking reactions were carried out in several mixing ratio condition. Figure 2 represents M_w data of reaction products after heating diblock copolymers **1** and **2** in several mixing ratio condition. The Job's plots show convex upward, which indicated that two types of diblock copolymers were reacted in complementarily reactive systems, because of controlling the directions of alkoxyamine units. The peak top seems to be reached between ([**1**]/[**1**]+[**2**]) =) 0.3 and 0.4, because the number of two types of alkoxyamine units become equalized at this mixing ratio. Consequently, the present system ensures to form star-like nanogels necessarily with two types of molecular chains as PSt and PMMA.

4. Conclusion

The authors have demonstrated the formation of hetero-arm star-like nanogels via cross-linking reaction of diblock copolymers with alkoxyamine units connected at different positions. The diblock copolymers with alkoxyamine units were prepared by ATRP methods, and GPC curves and M_w data showed two types of diblock copolymers were reacted in complementarily reactive system. The prepared star-like nanogels should have two types of chains as PSt and PMMA, and to the best of our knowledge, this is the first report on the synthesis of star polymers with different arms by using block copolymers. The present reaction system can be easily applied to core-functional star polymers, amphiphilic star-polymers, and so on.

Acknowledgements

The authors gratefully acknowledge the financial supports by a Grant-in-Aid for Scientific Research (20350057) and a Grant-in-Aid for the Global COE Program, "Science for Future Molecular Systems" from the Ministry of Education, Culture, Science, Sports and Technology of Japan. Y. A. acknowledges the financial support of a Grant-in-Aid for JSPS Fellows.

References

[1] Terashima T, Kamigaito M, Baek K.-Y., Ando T and Sawamoto M 2003 *J. Am. Chem. Soc.* **125** 5288.
[2] Iatrou H and Hadjichristidis N 1992 *Macromolecules* **25** 4649.
[3] Iatrou H and Hadjichristidis N 1993 *Macromolecules* **26** 2479.
[4] Kanaoka S, Omura T, Sawamoto M and Higashimura T 1992 *Macromolecules* **25** 6407.
[5] Zhao Y, Higashihara T, Sugiyama K and Hirao A 2007 *Macromolecules* **40** 228.
[6] Amamoto Y, Higaki Y, Matsuda Y, Otsuka H and Takahara A 2007 *J. Am. Chem. Soc.* **129** 13298.
[7] Otsuka H, Aotani K, Higaki Y and Takahara A 2002 *Chem. Commun.* 2838.
[8] Otsuka H, Aotani K, Higaki Y and Takahara A 2003 *J. Am. Chem. Soc.* **125** 4064.
[9] Rowan S J, Cantrill S J, Cousins G R L, Sanders J K M and Stoddart J F 2002 *Angew. Chem. Int. Ed.* **41** 898.
[10] Higaki Y, Otsuka H and Takahara A 2006 *Macromolecules* **39** 2121.
[11] Amamoto Y, Higaki Y, Matsuda Y, Otsuka H and Takahara A 2007 *Chem. Lett.* **36** 774.
[12] Amamoto Y, Kikuchi M, Maeda T, Otsuka H and Takahara A 2009 *Chem. Commun.* 689.
[13] Shipp D A, Wang J.-L. and Matyjaszewski 1998 *Macromolecules* **31** 8005.

Journal of Physics: Conference Series **184** (2009) 012020 doi:10.1088/1742-6596/184/1/012020

Tear energy and strain-induced crystallization of natural rubber/styrene-butadiene rubber blend

F Noguchi,[1] K Akabori,[1,2] Y Yamamoto,[1,2] T Kawazura[3] and S Kawahara[1,2,4]

[1]Department of Materials Science and Technology, Faculty of Engineering, Nagaoka University of Technology, Nagaoka 940-2188, Japan,
[2]Center for Green-Tech Development in Asia,
[3]Yokohama Rubber Company, Limited, 2-1, Oiwake, Hiratsuka Kanagawa, 245-8601,Japan

E-mail: kawahara@mst.nagaokaut.ac.jp

Abstract. Strain-induced crystallization of natural rubber (NR), dispersed in styrene-butadiene rubber (SBR), was investigated in relation to dimensional feature of a dispersoid and crosslink density of NR by measuring tear energy (*G*) of crosslinked NR/SBR blends. The crosslinked NR/SBR blends in ratios of 1/9 and 3/7 by weight were prepared by mixing masticated NR and SBR with an internal mixer at a rotor speed of 30 rpm, followed by crosslinking with dicumyl peroxide on a hot press at 444 K for 60 min. The *G*, measured in wide-ranges of temperature and tear rate, was superposed into a master curve with a Williams-Landel-Ferry shift factor. The *G* of the NR/SBR(3/7) blend abruptly decreased to a level comparable to that of SBR at about melting temperature of NR crystals formed on straining. The temperature, at which the dramatic decrease in the *G* occurred, was associated with the dimensional feature of the NR dispersoid and the crosslink density.

1. Introduction

Crystallization of a minor component as a dispersoid is of great importance to control mechanical properties of rubbery multi-component system, i.e. rubber blend. It has been anticipated that the crystallization of the minor component such as a crystallizable rubber improves tensile strength and tear strength of the major amorphous rubber as a matrix. However, little work has been studied on an effect of the crystallization of the minor crystallizable rubber on the properties of the major amorphous rubber. Thus, it is necessary to investigate the crystallization of the minor crystallizable rubber in the rubber blend and its effect on the mechanical properties of the major amorphous rubber.

Finely-divided material is widely recognized to crystallize more slowly because a probability of finding a suitable crystal nucleus in small volumes of the materials is quite low. For example, the crystallization of natural rubber (NR) in latex, the particle size of which was less than 1 μm in diameter, was suppressed at -29 °C by a factor of at least 10 times in comparison with the bulk material.[1] In contrast, our previous study demonstrated that the rapid crystallization due to heterogeneous nucleation occurred at -25 °C when the average diameter of the NR particle was longer than about 1 μm.[2] These suggest that the crystallization of NR is dependent upon the dimensional

[4] Author to whom all any correspondence should be addressed.

feature of the particle. It is, thus, interesting to investigate a relationship between the particle dimension and the mechanical properties, since the outstanding mechanical properties are closely related to the strain-induced crystallization.

Recently, we proposed a novel method[3] to investigate the strain-induced crystallization of NR, which was a measurement of tear energy (*G*) of the NR/styrene-butadiene rubber (SBR) blend, since the strain-induced crystallization plays an important role in preventing crack growth of the blend under large deformation.[4] The *G* of the NR/SBR blend was superposed into two curves with a Williams-Landel-Ferry (WLF) shift factor, whereas the *G* was superposed into a single master curve, as in the previous works.[5] The difference in the *G* between the NR/SBR blend and the SBR was assumed to be due to the strain-induced crystallization of NR.

In this study, the effects of dimensional feature and crosslink density on the *G* of the crosslinked NR/SBR blend were investigated to prove the strain-induced crystallization of NR in the blend. The NR/SBR blend was crosslinked with dicumyl peroxide. The morphology of the blend was observed by transmission electron microscopy. The principle of time-temperature superposition was applied to make the master curve of the *G* of the blend over wide ranges of temperature and tear rate.

2. Experimental

NR used in this study was prepared by coagulation of commercial high-ammonia natural rubber latex (Golden hope) followed by drying at 50 °C for a week under reduced pressure. SBR was kindly supplied from Nippon Zeon Co., Ltd. (Nipol1502). NR was masticated with an internal mixer (TOYO SEIKI SEISAKU-SYO, Ltd.) at a rotor speed of 30 rpm for 15 min. The masticated NR was mixed with SBR in ratios of 3/7 (NR/SBR(3/7) blend) and 1/9 (NR/SBR(1/9) blend) by weight for 20 min. Dicumyl peroxide (DCP, PERCUMYL D-40(NOF CORPORATION)) was added to the blends, and then they were cured on a hot press at 444 K for 60 min. The crosslink density was estimated in terms of the Flory-Rehner equation[6] from swelling ratios determined with benzene and *n*-heptane, respectively.[7]

Morphological observation for the NR/SBR(3/7) and NR/SBR(1/9) blends was performed with a JEOL (Tokyo, Japan) JEM-2100 at an accelerating voltage of 200 kV. The ultrathin section of the blends was prepared with an ultramicrotome equipped with a REICHERT-NISSEI ULTRACUT N below 200 K. The section was stained with RuO_4.

The tear strength (*G*) was measured with trousers-type strips at various temperatures and tear rates. A specimen used was prepared according to the previous method.[3] The tear rate (*R*) was defined as a mean value for the crosshead speed of the testing machine.

Table 1. Crosslink density of NR/SBR blends.

Sample	Crosslink density $\times 10^{-4}$(mol cm^{-3})	
	NR phase	SBR phase
NR/SBR(3/7)	0.913	1.24
NR/SBR(1/9)	0.826	1.02
H-NR/SBR(3/7)	2.76	3.69

3. Results and Discussion

Crosslink density of NR/SBR blends, estimated by the following Flory-Rehner expression,[6] is tabulated in Table 1:

$$\nu = \frac{S_B + \chi_B S_B{}^2 + \ln(1 - S_B)}{V_B(S_B{}^{1/3} - S_B/2)} = \frac{S_H + \chi_H S_H{}^2 + \ln(1 - S_H)}{V_H(S_H{}^{1/3} - S_H/2)}$$

where ν, *S*, *V* are the crosslink density of the rubber, the volume fraction of the rubber in the swollen specimen, and the molar volume of the solvent, respectively; the subscripts represent the solvents used, i.e., benzene and *n*-heptane. The values of interaction parameters (χ) between the rubber and the solvents were referred to the literatures.[8,9] The estimated value of the crosslink density of the NR-phase in each blend was almost similar to that of the SBR phase. The value was independent of the composition of the blend; that is, the value of the NR/SBR(3/7) blend was identical to that of the NR/SBR(1/9) blend. The value of the H-NR/SBR(3/7) blend was about 3 times as large as that of the NR/SBR(3/7) blend.

Figure 1 (a) and (b) show TEM images for NR/SBR(1/9) blend and NR/SBR(3/7) blend, respectively. The bright domain is corresponding to NR-rich phase and dark domain is to SBR-rich phase, since the blends were stained with RuO_4. As for the NR/SBR(1/9) blend, the NR domain of less than 1 μm in diameter was dispersed into the SBR matrix, in which the geometrical feature of the NR-domain as a dispersoid was spherical, as in the case of ordinary immiscible polymer blends. In contrast, for the NR/SBR(3/7) blend, the feature of the NR dispersoid was ellipsoidal rather than spherical and its diameter was much longer than that of the NR/SBR(1/9) blend. The ellipsoidal feature of the NR domains may be explained to be due to an unsatisfied mechanical mixing.

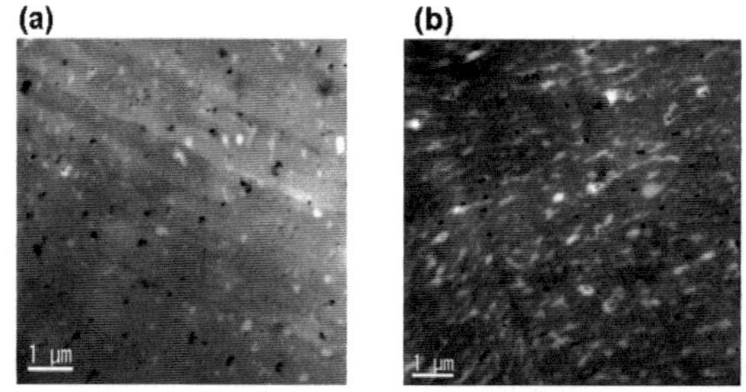

Figure 1. Typical TEM images of the blends of (a) NR/SBR(1/9) and (b) NR/SBR(3/7).

Figure 2 shows the *G* of NR/SBR(3/7) and NR/SBR(1/9) blends, which are represented as filled and open symbols, respectively. The solid line corresponds to a master curve of *G* of SBR superposed with a shift factor. The *G* of the NR/SBR(1/9) blend was in good agreement with that of SBR. In contrast, the *G* of the NR/SBR(3/7) blend was larger than that of SBR itself, represented as a solid line. Furthermore, the *G* of the NR/SBR(3/7) blend was divided into two curves: one for the *G* measured at 273 and 278 K and the other at temperatures ranging from 298 to 333 K, whereas the *G* of the NR/SBR(1/9) blend was not; i.e. the single master curve. The abrupt increase in the *G* of the NR/SBR(3/7) blend may be attributed to a reinforcing effect of the strain-induced crystals of NR dispersoid as a filler.

The effect of the crosslink density of NR on the *G* of the NR/SBR blend was investigated with the NR/SBR(3/7) and H-NR/SBR(3/7) blends. Figure 3 shows the master curves of the *G* of the NR/SBR(3/7) and H-NR/SBR(3/7) blends, which are represented as open and filled symbols, respectively. The *G* of the H-NR/SBR(3/7) blend was well-superposed to each other to make a single master curve, which was distinguished from the *G* of the NR/SBR(3/7) blend. This demonstrates that the strain-induced crystallization of NR dispersoid is suppressed by crosslinking, since the crosslink density of H-NBR/SBR(3/7) blend was higher than that of NR/SBR(3/7) by a factor of 3. Consequently, the *G* of the NR/SBR blend was found to be dependent upon the strain-induced crystallization, which was concerned with not only morphology of the blend but also the crosslink density of NR.

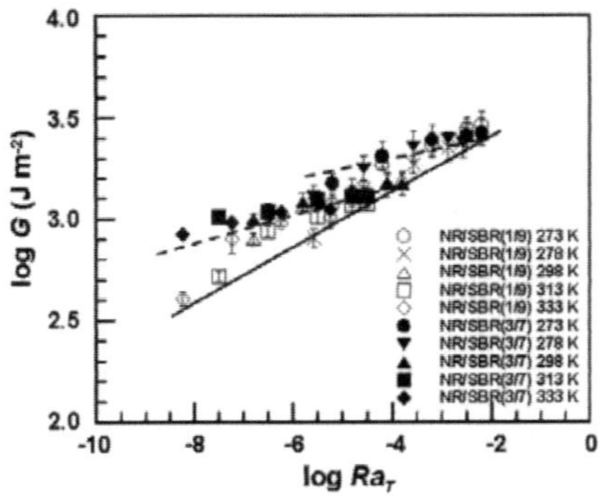

Figure 2. The master curve after the superposition of G for the NR/SBR(3/7) and NR/SBR(1/9) blends.

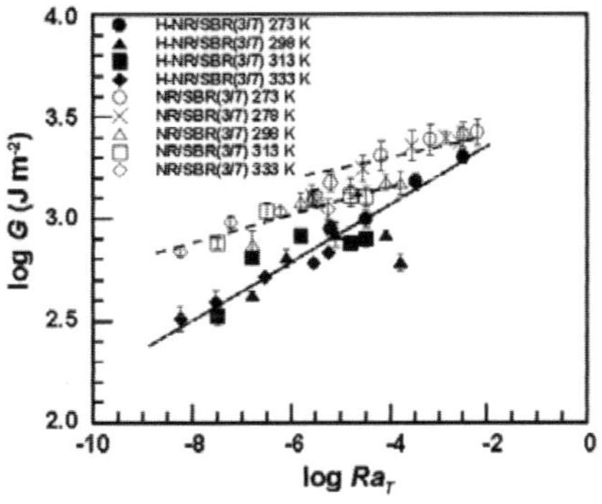

Figure 3. The master curve after the superposition of G for the NR/SBR(3/7) and H-NR/SBR(3/7) blends.

4. Conclusions

The geometrical feature of the NR domain in the NR/SBR(3/7) blend was ellipsoidal rather than spherical and its average diameter was longer than about 1 μm, whereas the spherical domain of less than 1 μm in diameter was dispersed in the NR/SBR(1/9) blend. The G of the NR/SBR(3/7) blend was divided into two curves after superposition: one for the G measured at 273 K and the other at temperatures ranging from 298 to 333 K, though that of the NR/SBR(1/9) blend was superposed into the single master curve, as in the case of SBR. When the crosslink density of NR increased by a factor of 3, the abrupt increase in the G for the NR/SBR(3/7) blend disappeared, suggesting that the increase in the crosslink density of NR suppresses the strain-induced crystallization. Thus, it is concluded that the strain-induced crystallization is controlled by not only the dimensional feature of the NR domain but also the crosslink density of NR.

References

[1] Wathen G D and Gent A N 1990 *J. Nat. Rubber Res.* **5** 178
[2] Kawazura T, Kawahara S and Isono Y 2003 *Rubber. Chem. Technol.* **76** 1164
[3] Kawazura T, Kawahara S and Isono Y 2005 *J. Appl. Polym. Sci.* **98** 613
[4] Andrews E H 1962 *Rubber Chem. Technol.* **35** 210
[5] Greensmith H W and Thomas A G 1955 *J. Polym. Sci.* **18** 189
[6] Flory P J and Rehner J 1943 *J. Chem. Phys.* **11** 521
[7] Wang, Y F and Wang C 1997 *Rubber. Chem. Technol.* **70** 663
[8] Bristow G M and Watson W F 1958 *Trans. Faraday Soc.* **54** 1731
[9] Tewari Y B and Schreiber H P 1972 *Macromolecules* **5** 329

Journal of Physics: Conference Series **184** (2009) 012021 doi:10.1088/1742-6596/184/1/012021

Coordination Ability of Poly(vinylene-arsine)s towards Transition Metal Ions

A Nakahashi[1], M Bravo[1], K Naka[2], Y Chujo[1]

[1]Department of Polymer Chemistry, Graduate School of Engineering, Kyoto University, Katsura, Nishikyo-ku, Kyoto, 615-8510, Japan
[2]Department of Chemistry and Materials Technology, Graduate School of Scienece and Technology, Kyoto Institute of Technology, Matsugasaki, Sakyo-ku, Kyoto, 606-8585, Japan

E-mail: naka@chem.kit.ac.jp

Abstract. Coordinations of poly(vinylene-arsine)s (**1**) towards transition metal ions were carried out to estimate the coordination ability of phenyl-substituted poly(vinylene-arsine) (**1a**) and methyl-substituted poly(vinylene-arsine) (**1b**). No **1a**-Co complex was obtained upon addition of an acetone solution of **1a** to an acetone solution of cobalt (II) chloride hexahydrate. On the other hand, **1b**-Co complexes were obtained in the reactions at different cobalt salt-**1b** feed ratios. As in the usual case, the bivalent complexes of the reaction run corresponding to the feed ratio of 1:1 [$CoCl_2$]/[repeating unit of **1b**] oxidized readily in air to give very stable complexes of trivalent cobalt. On the other hand, the bivalent complexes of the reaction runs corresponding to the feed ratios of 1:4 and 1:10 both exhibited unusual stability against air oxidation owing to the higher concentration of **1b** in the reaction mixture. Reactions of **1** and bis(2,2'-bipyridyl)ruthenium(II) dichloride also provided polymer complexes. In all the cases, **1b** had higher coordination property than **1a** due to its lower steric hindrance.

1. Introduction

The introduction of transition metal complexes into conjugated polymer backbones provide new dimensions of novel materials with improved electrical, magnetic and catalytic properties. Moreover, the polymer metal complexes offer advantages like recovering from reaction mixtures by simple filtration or reprecipitation, and easy recycling. A variety of transition metal polymer complexes have been synthesized, in which the metal complexes are linked to the backbones as isolated pendants, or are inserted into the polymer backbones.

Poly(vinylene-arsine)s (**1**) are expected to form polymer complexes with various kinds of transition metals due to the lone pair on the arsenic atom. Because **1** have arsenic atoms in the backbone, the transition metal can coordinate directly towards its backbone, and the connection of transition metals can be expected. In some reports, organoarsenic compounds have been reported to be superior ligands [1-4]. For example, osmium complexes with the arsenic ligands exhibited phosphorescence with greater quantum yields. However, only one polymer complex coordinated with arsine has been reported [5]. The reason for this is originally difficult synthesis of organoarsenic compounds due to the toxicity of the arsenic sources (arsenic chloride or arsenic hydride, for example). As shown in the previous works, poly(vinylene-arsine)s (**1**) have been synthesized by ring-collapsed radical alternating copolymerization (RCRAC) of phenylacetylene with hexaphenylcyclohexaarsine (*cyclo*-$(PhAs)_6$) or pentamethylcyclopentaarsine (*cyclo*-$(MeAs)_5$), which is the first soluble polymers containing arsenic atoms in the main chain (Scheme 1) [6,7]. In this paper, the coordination abilities of poly(vinylene-arsine)s (**1**) towards Co ions and Ru ions are estimated.

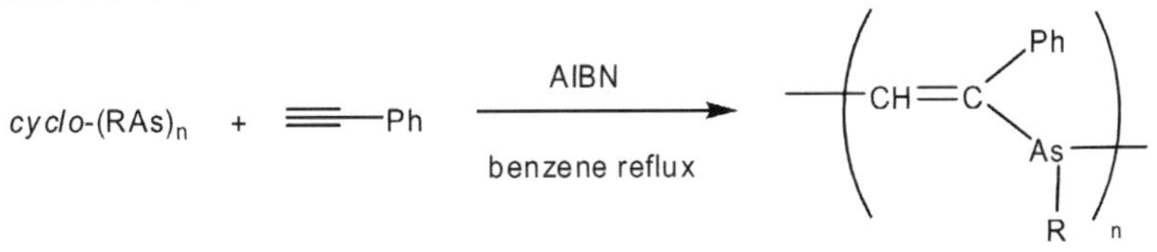

Scheme 1. Preparation of poly(vinylene-arsine)s (**1**).

2. Experimental

2.1. Materials

Unless otherwise noted, all reagents and chemicals were purchased and used without further purification. Hexaphenylcyclohexaarsine (*cyclo*-$(PhAs)_6$) [8] and pentamethylcyclopentaarsine (*cyclo*-$(MeAs)_5$) [9] were prepared according to literature procedures. Phenylacetylene was distilled under reduced pressure. Poly(vinylene arsine)s (**1**) were prepared by the method mentioned in other reports [6,7].

2.2. Instruments

All reactions were carried out under an argon atmosphere using Schlenk techniques. Gel permeation chromatography was carried out on a TOSOH UV-8011 and RI-8000 (Shodex K-803L column) using $CHCl_3$ as an eluent after calibration with standard polystyrene. ^1H and ^{13}C NMR spectrum was obtained using a JEOL JNM-EX270 instrument (270 and 67.5 MHz, respectively) for solutions in $CDCl_3$ or CD_3CN and were referenced to $SiMe_4$ (TMS). UV-vis spectra were measured on a JASCO V-530 spectrometer, and the samples were analyzed in $CHCl_3$ at room temperature and on a Perkin Elmer Lambda 19 spectrometer for powder samples analyzed at room temperature. Fourier transform infrared (FT-IR) spectra were obtained on a Perkin Elmer 2000 spectrometer. Cyclic voltammetry analysis was done with a BAS CV-50W Electrochemical Analyzer in a CH_2Cl_2 solution of 0.1M $[CH_3(CH_2)_3]_4N\text{-}PF_6$ as a supporting electrolyte. Platinum wire auxiliary electrode and Ag/AgCl RE-5 reference electrode were used in the CV measurement.

2.3. Preparation of Poly(vinylene-arsine)s-Cobalt Complex.

Cobalt chloride hexahydrate (12 mg) in acetone (2 ml) was treated with **1b** (10 mg) in acetone (4 ml) (Run 6). The reaction mixture was stirred at room temperature for 1 d, heated on the oil-bath under reflux condition for 30 min, and then cooled. The yellow-green complex was isolated by filtration, and then well washed with hexane and acetone. The complex was finally dried under vacuum (product weight = 9 mg). On the other hand, the solvent was removed from the filtrate under reduced pressure and the residue consisting of the uncoordinated cobalt ions and **1b**-Co complex was collected. H_2O (25 ml) was then used to dissolve the uncoordinated cobalt ions and isolate them from **1b**-Co complex. This uncoordinated cobalt solution (aq.) was used in the determination of the amount of uncoordinated Co ions by EDTA titration (the detailed procedure is presented below).

2.4. Determination of Uncoordinated Cobalt Ions in Aqueous Solution: Direct Titration Using Xylenol Orange Indicator

The uncoordinated cobalt aqueous solution (25 ml) was pipetted into a 500 ml conical flask and diluted with de-ionized water to about 150 ml. The pH was then adjusted to about 6 by the cautious addition of hexamine (pH meter was used). Then, 30 mg of the Xylenol Orange/H_2O solution was added and then titrated with standard 0.01*M*-EDTA solution until the purple colour started to fade.

From this point, the titrant was added slowly until the end-point was reached and the indicator changed to yellow (1 mole EDTA ≡ 1 mole Co^{2+}) [10].

2.5. Preparation of Poly(vinylene-arsine)s-Ruthenium Complex.
A typical experimental procedure is as follows. A half equimolar amount (vs vinylene-arsine structure) of *cis*-$(bpy)_2RuCl_2 \cdot 2H_2O$ and poly(vinylene-arsine) were stirred in a refluxing deaerated $CHCl_3$ / ethanol (1/1) for 24 h. The solvent of the resulting mixture was reduced by evaporation and the resulting solution was poured into methanol solution of NH_4PF_6. After reprecipitation several times, the polymer complex was obtained as a pink-orange powder (**2**).

3. Results and Discussion

3.1. Poly(vinylene-arsine)-Cobalt Complex
Poly(vinylene-arsine)-cobalt complexes have been prepared and studied because the coordination of the compounds of tertiary arsines with cobalt has been well studied providing background information for this research [11-14].

Table 1. Results of polymerization

run	Feed ratio [$CoCl_2$]/[**1b**]	Reaction condition **1b** (mg)	acetone (ml)	**1b**-Co complex	Colour of the complex
1	1:20	200	40	79	Blue-green
2	1:10	100	40	42	Blue-green
3	1:1	10	40	4	Yellow-green
4	1:10	100	30	37	Blue-green
5	1:4	40	15	28	Sky blue
6	1:1	10	4	9	Yellow-green

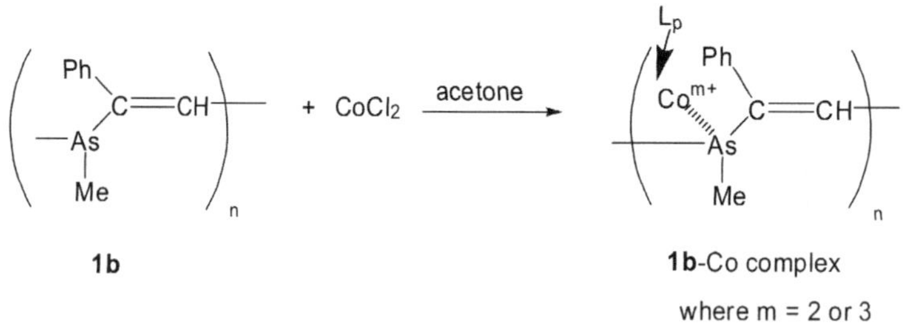

Scheme 2. Preparation of cobalt–complex (**1**-Co).

Polymer metal complexes were prepared by adding an acetone solution of **1a** or **1b** to an acetone solution of cobalt chloride hexahydrate at different feed ratios. The reaction mixtures were stirred for 1 day at room temperature, refluxed for 30 min, and then, cooled. No **1a**-Co complex was obtained due to the bulky structure (phenyl group) of **1a**. On the other hand, **1b**-Co complexes precipitated even before refluxing (Scheme 2). The products were isolated by filtration. The residue was then washed with *n*-hexane and acetone, and finally, dried under vacuum. The results are summarized in Table 1.

The **1b**-cobalt complexes exhibit different colors. The color of the complex is some guide to the stereochemistry. Mellor and Craig reported that square complexes of bivalent cobalt were some shade of brown while the ionic or tetrahedral complexes varied in color from pink or mauve to blue. In this case, the blue-green complexes of run 1 corresponding to the feed ratio of 1:20 [$CoCl_2$]/[repeating unit of **1b**] and runs 2 and 4 corresponding to the feed ratio of 1:10, and the sky-blue complexes of run 5 corresponding to the feed ratio of 1:4 suggest complexes of bivalent cobalt.

Moist cobalt(II) chloride becomes green on the surface in air, owing to oxidation, and bivalent complexes are oxidized readily in air to give very stable complexes of trivalent cobalt. When air is admitted to the reaction mixture, oxidation takes place rapidly and the cobalt(II) salt dissolves to give a green solution from which a green complex of trivalent cobalt was obtained on cooling [14]. This explains the yellow-green complexes of runs 3 and 6 corresponding to the feed ratio of 1:1. Apparently, the bivalent complexes of runs 3 and 6 were sensitive to oxidation by air due to the lower polymer concentration in the reaction mixture.

It has been established that the colors of the **1b**-Co complexes which serve as some guide to the stereochemistry of the complexes and valency of the coordinated cobalt could be controlled by the feed ratio of cobalt(II) chloride hexahydrate and **1b**. This is the particularly interesting result since it may mean controlling the chemical and physical properties of the complexes for specific applications.

Cobalt(II) chloride formed stable, slightly soluble complexes with **1b**. The complexes of runs 4, 5, and 6 corresponding to the feed ratios of 1:10, 1:4, and 1:1 were analyzed by FTIR and were observed by their colors before and after washing with *n*-hexane and acetone to investigate their stability against oxidation. No changes were observed in the IR spectra and in the colors of the complexes before and after washing showing that the complexes are indeed stable. The cobalt salt was insoluble in chloroform and **1b** was soluble. While the obtained polymer metal complexes of runs 1 and 4 are soluble and those of run 6 and of the other runs are slightly soluble in the same solvent. All the obtained complexes are only either slightly soluble or insoluble in other solvents such as DMF, DMSO, benzene, acetone, and THF, which dissolves polymer **1b**.

To estimate the Co ion content in the polymer metal complex, the amount of uncoordinated Co ions was determined by EDTA titration. The molar ratios of Co ion and repeating unit of **1b** in the polymer metal complexes of runs 4, 5, and 6 are 1:7, 1:3, and 1:1, respectively. Complexes of runs 4 and 5 corresponding to the feed ratio of 1:10 and 1:4, respectively, are expected to be complexes of bivalent cobalt as exhibited by their colors. The reluctance of the bivalent cobalt to form octahedral complexes with covalent bonds is well known [14]. This is consistent with the estimated molar ratios of Co ion and repeating unit of **1b** in the polymer metal complexes of runs 4 and 5. For run 6, the estimated molar ratio of Co ion and repeating unit of **1b** in the polymer metal complex is reasonable owing to the steric hindrance involved in the coordination of the polymer towards the metal ion.

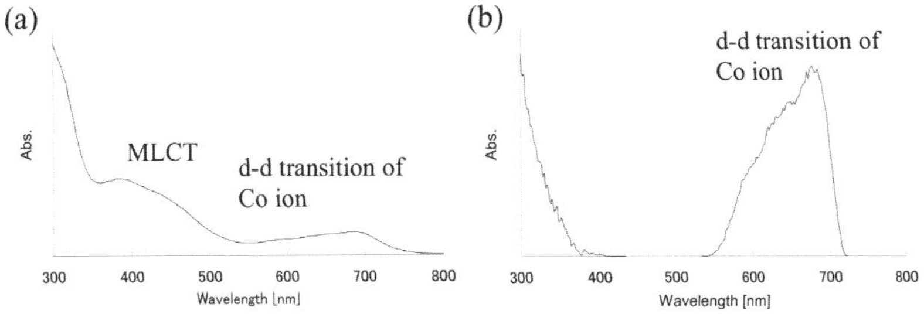

Figure 2. UV-vis absorption spectra of (a) poly(vinylene-arsine)-cobalt complexes of run 6 (1:1 [$CoCl_2$]/[repeating unit of **1b**] feed ratio) in powder, (b) poly(vinylene-arsine)-cobalt complexes of run 4 (1:10 feed ratio) in powder.

The UV-vis spectrum of the complex of run 6, which corresponds to the feed ratio of 1:1, showed a metal-to-ligand charge transfer (MLCT) at 370 nm (Figure 2a). This absorption suggests that the bivalent complexes of this reaction run oxidized in air to give very stable complexes of trivalent cobalt. On the other hand, no MLCT is observed in the UV-vis spectra of the complexes of run 4, which corresponds to the feed ratio of 1:10 (Figure 2b) and of the complexes of run 1, which corresponds to the feed ratio of 1:20, in chloroform showing that the bivalent complexes of these reaction runs exhibit unusual stability against air oxidation owing to the higher concentration of **1b** in the reaction mixture.

3.2. Poly(vinylene-arsine)- Ruthenium Complex

Complexations of the poly(vinylene-arsine)s with ruthenium ions were carried out (Scheme 3). Half equimolar amount (vs vinylene-arsine structure) of *cis*-$(bpy)_2RuCl_2{\cdot}2H_2O$ and the poly(vinylene-arsine) are stirred in a refluxing deaerated $CHCl_3$ / ethanol (1/1). The solvent of the resulting mixture was reduced by evapolation and the resulting solution was poured into methanol solution of NH_4PF_6 to yield the polymer complex. The polymer was purified by repeating the reprecipitation. In the case of phenyl-substituted poly(vinylene-arsine) (**1a**), the obtained polymer complex (**2a**) was little colorized, which suggests little amount of Ru ions. On the other hand, the complexation of methyl-substituted poly(vinylene-arsine) (**1b**) resulted in red solid (**2b**). The obtained polymer **2b** was soluble in CH_2Cl_2 and acetonitlile, and insoluble in benzene. The ^1H NMR spectrum of **2b** showed the peaks assignable to the bipyridyl group. The new peaks of the methyl protons (a') appeared at δ 0.3 ppm — −0.8 ppm, which are derived from the coordination of ruthenium ions. The integral ratio of the peak attributed to the aromatic and vinyl protons and the peak of the methyl protons was 8.94 : 3 and that of the peak (a') and the sum of the methyl peaks was 0.58 : 3. The estimations from these ratio suggested that 18.3% and 19.4% of arsines were coordinated with ruthenium ions, respectively. On the other hand, there was little difference in the ^1H NMR spectra of **2a** and **1a**, which suggests no or little complexation with ruthenium ions.

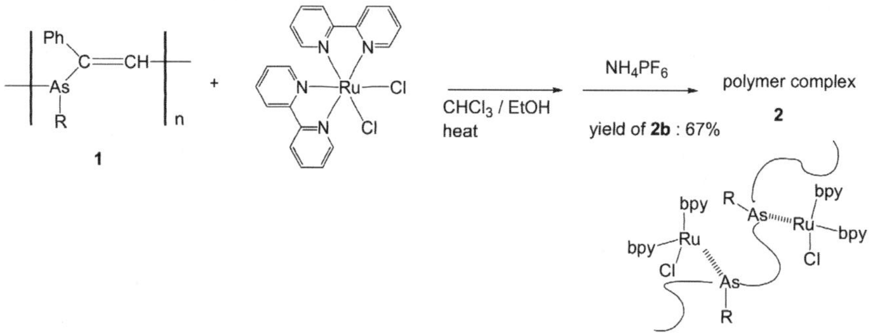

Scheme 3. Preparation of ruthenium–complex (**2**).

The UV-vis absorption spectra of **2a** and **2b** showed an absorption peak at around 470 nm, which is derived from metal-to-ligand (bpy) charge transfer (MLCT) as well as an strong absorption due to the aromatic groups in the UV region. This result also indicates the incorporation of the complexes. Cyclic voltammetry (CV) measurement of **2a** in CH_2Cl_2 showed a reversible oxidation peak derived from the oxidation of Ru(II) ions. The potential difference between the oxidation peak and the half oxidation peak was 40 mV, which suggests little interaction between Ru ions when around 20% of arsine is coordinated to Ru ions.

4. Conclusions

The coordination ability the poly(vinylene-arsine)s towards Co ions and Ru ions in acetone were described. The poly(vinylene-arsine) derived from hexaphenylcyclohexaarsine (**1a**) formed no cobalt complex upon addition of an acetone solution of the polymer to an acetone solution of cobalt(II) chloride hexahydrate due to its bulky structure. On the other hand, the poly(vinylene-arsine) derived from pentamethylcyclopentaarsine (**1b**) formed the **1b**-Co complexes at different cobalt salt-polymer ratios due to its less bulky structure. As in the usual case, the bivalent complexes of the reaction run corresponding to the feed ratio of 1:1 [$CoCl_2$]/[repeating unit of **1b**] oxidized readily in air to give very stable complexes of trivalent cobalt. On the other hand, the bivalent complexes of the reaction runs corresponding to the feed ratios of 1:10 and 1:4 both exhibited unusual stability against air oxidation owing to the higher concentration of the polymer in the reaction mixture. The complexations of the poly(vinylene-arsine)s with ruthenium ions were carried out to result in the lower coordination ability of **1a**. The obtained polymer complex showed the absorption peak derived from the incorporated ruthenium complex. The polymer also exhibited a reversible oxidation peak of Ru ion.

References

[1] Baber R A, Collad S, Hooper M, Orpen A G, Pringle P G, Willkinson M J and Wingad R L 2005 *Dalton Trans.* **124** 1491
[2] Farina V and Krishman B 1991 *J. Am. Chem. Soc.* **113** 9585
[3] Jeanneret V, Meerpoel L and Vogel P 1997 *Tetrahedron Lett.* **38** 543
[4] Carison B, Phelan G D, Kaminsky W, Dalton L, Jiang X, Liu S and Jen A K Y 2002 *J. Am. Chem. Soc.* **124** 14162
[5] Cai M, Huang Y, Zhao H and Song C 2003 *J. Organomet. Chem.* **682** 20
[6] Naka K, Umeyama T and Chujo Y 2002 *J. Am. Chem. Soc.* **124** 6600
[7] Naka K 2008 *Polym. J.* **40** 1031
[8] Elmes P S, Middleton S and West B O 1970 *Aust. J. Chem.* **23** 1559
[9] Reesor J W B and Wright G F 1957 *J. Org. Chem.* **22** 382
[10] Bassett J 1978 *Vogel's Textbook of Quantitative Inorganic Analysis: Including Elementary Instrumental Analysis* 4th edition (London ; New York : Longman)
[11] McAuliffe C A Levason W 1979 *Phosphine, Arsine, and Stibine Complexes of the Transition Elements* (Elsevier Scientific Pub. Co., Amsterdam, New York)
[12] Feltham R D Metzger H G and Silverthorn W 1968 *Inorg. Chem.* **7** 2003
[13] Feltham R Dand Silverthorn W 1968 *Inorg. Chem.* **7** 1154
[14] Nyholm R S 1950 *J. Chem. Soc.* 2071

Journal of Physics: Conference Series **184** (2009) 012022 doi:10.1088/1742-6596/184/1/012022

Thermal Deformation of Imprinted Twist Nematic Elastomers

Y Sawa, K Urayama[1], and T Takigawa

Department of Materials Chemistry, Kyoto University, Nishikyo-ku, Kyoto 615-8510, Japan

E-mail: urayama@rheogate.polym.kyoto-u.ac.jp

Abstract. A nematic elastomer film with a 90° twist orientation between the top and bottom surfaces is prepared by chiral imprinting methods. The reactive achiral nematic monomers and crosslinkers are photopolymerized in the presence of a controlled amount of unreactive chiral dopant. The elastomer films possess a 90° twist orientation even after the removal of the chiral dopant. The films exhibit an elongation and contraction in the direction at an angle of 45° relative to the mesogen orientations at the top and bottom surfaces in response to cooling and heating across the nematic–isotropic transition temperature, respectively.

1. Introduction

A marked characteristic of liquid crystal elastomers (LCEs) is the strong coupling of macroscopic shape and the molecular (mesogen) orientation [1, 2]. One of the most familiar phenomena in LCEs is the large distortions induced by changes in temperature. In the case of nematic elastomers, variations in temperature around the nematic–isotropic transition temperature (T_{NI}) lead to uniaxial deformation; this deformation results from the change between uniaxial and random orientations of the constituent mesogens [1, 3–6]. In the case of smectic elastomers, shear deformation occurs when a tilt state of mesogens in the smectic layer changes due to the transition between the smectic A and smectic C phases [7]. These results clearly indicate that the mode of thermal deformation in LCEs is governed by a thermally induced change in the patterns of mesogen orientation. In this study, we focus on nematic elastomers with a 90° twist orientation between the top and bottom surfaces. The twist nematic (TN) orientation of low molecular mass LCs has been exploited in display devices [8]. Then, the following question arises: what is the thermal deformation of twist nematic elastomers (TNEs) like? This is an interesting issue in view of the coupling effects of mesogen alignment and rubber elasticity in LCEs. In addition, an understanding of the thermal deformation of TNEs will enable us to extend the possibility of the application of LCEs as thermally driven soft actuators [9].

"Chiral imprinting" provides a method of producing TNEs. The concept of imprinted chiral elastomers was envisaged by de Gennes [10], and chiral imprinting was experimentally attempted to produce achiral nonmesomorphic [11] and LC polymers [12–16]. Hassan et al. [13] first reported that when achiral LC polymers are cross-linked in the presence of a nonreactive chiral dopant (low molecular weight chiral mesogen), the resultant LCEs exhibit an imprinted chiral structure. The LCEs retain the induced chiral nematic phase after cross-linking even if the chiral dopant is removed. The physical properties of cholesteric elastomers prepared by chiral imprinting have been experimentally [14–16] and theoretically [17] studied. These cholesteric elastomers have a helical orientation with any

[1] To whom correspondence should be addressed.

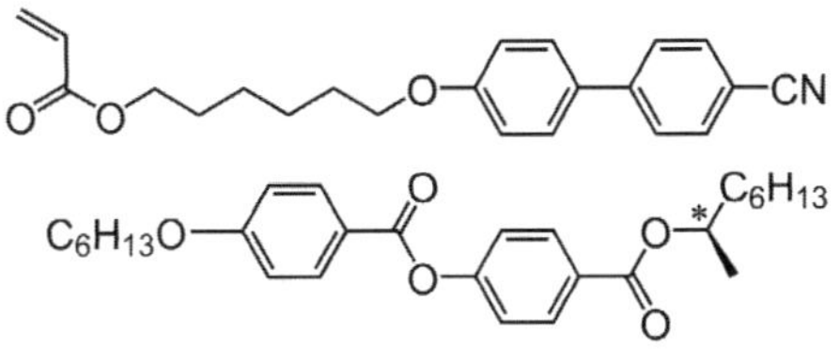

A-6OCB

S-811

Figure 1. Chemical structures of A-6OCB and S-811.

windings. Broer et al. [12] prepared LC network films with a twist orientation by using the chiral imprinting method. These films –which were prepared by the homopolymerization of achiral mesogenic diacrylate with low molecular weight– were not elastomers but hard resins because of the considerably short network chains. They observed the interesting thermal deformation of these films, but it originates from the difference in thermal expansion coefficient along the thickness direction and not from the director orientation.

2. Experimental

We have prepared a side-chain-type LCE with an imprinted 90° twist orientation using the achiral mesogenic monoacrylate (A–6OCB), 1,6-hexanediol diacrylate (cross-linker) and S-811 (nonreactive chiral dopant). The chemical structures of A-6OCB and S-811 are shown in Figure 1. 4-*n*-Hexyloxy-4'-cyanobiphenyl (6OCB) was used as a miscible nematic solvent to increase the temperature range of the nematic phase. A-6OCB was synthesized by the method specified in literature [18]. The chiral dopant and photoinitiator (Irgacure 784) were supplied by Merck Ltd., Japan, and Chiba Specialty Chemicals Co., respectively. 6OCB was purchased from Aldrich and used without further purification. The molar mixing ratio of A-6OCB and 6OCB was 1:1. The cross-linker concentration was 7 mol%.

The chiral properties of the mixtures of A-6OCB, 6OCB, and S-811 were characterized by the pitch of the resultant helicoidal structure corresponding to the distance of a 2π screw rotation of the molecular packing. The pitch was measured by the Cano wedge method [8, 19, 20]. The inverse pitch almost linearly depends on the concentration of S-811. This behavior is similar to the behaviors of conventional LC mixtures with chiral dopants. The total twist angle of the monomeric mesogens was defined as the angle between the bounding mesogenic molecules at the bottom and top substrates, whose surfaces were coated with a uniaxially rubbed polyimide layer. The chiral dopant induced a left-hand twist orientation of the mesogens. The top and bottom substrates of the glass cells used to prepare the sample were placed in such a manner that the rubbing directions could cross with each other. The concentration of the chiral dopant was adjusted such that the total twist angle could become 90° between the top and bottom substrates, the distance between the substrates being 20 μm. Photopolymerization was conducted in the cell at 45 °C—ca. 3 °C lower than the transition temperature of the nematic mixture—by irradiation for a duration of 30 min. A xenon lamp emitting at a wavelength of 526 nm was used. The cell was immersed in dichloromethane for several days until the resultant film detached from the glass substrates due to the swelling pressure. The films were allowed to swell fully in dichloromethane in order to remove the unreacted materials and chiral dopants. The swollen gel films were gradually deswollen by stepwise addition of methanol to dichloromethane. The specimens were dried and then used for the measurements. Square specimens with a thickness of ca. 15 μm—whose crossed sides were parallel to the rubbing directions at the top and bottom substrates—were cut out from the sheet sample at room temperature which was below T_{NI} (103 °C) and the glass transition temperature (ca. 35 °C) [21].

The phase behavior and thermal deformation of the TNEs were observed by means of a polarized optical microscope with a Mettler FP-82 hot stage. The mesogen orientation was characterized by an FTIR spectrometer Jasco 4200ST with a custom-made temperature-controllable box. The adsorption band for the stretching vibration of the terminal cyano group that was parallel to the long axis of the mesogen was used to characterize the mesogen orientation. The absorbance (A) of the unpolarized incident light normal to the film surface was measured as a function of temperature.

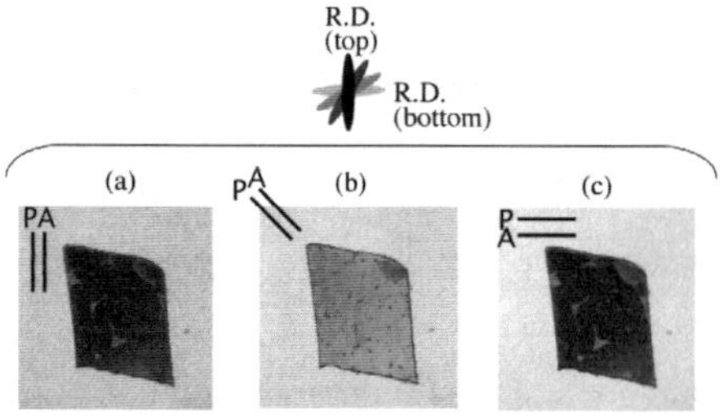

Figure 2. Optical micrographs of the specimen placed between two parallel polarizers at 93 °C (=T_{NI} – 10 °C). The optical axis of the parallel polarizers is (a) parallel and (c) normal to the rubbing direction (R.D.) for the top surface and (b) at an angle of 45° to the crossed rubbing directions.

3. Results and Discussion

Figure 2 shows the optical micrographs of the specimen. These micrographs confirm a 90° twist mesogen orientation. The sample is placed between two parallel polarizers. When the optical axis of the parallel polarizers coincides with the rubbing direction at the top or bottom surface, the sample appears dark. A bright image is observed when the optical axis of the polarizers is inclined 45° to the crossed rubbing directions. A similar twist orientation is observed at temperatures less than T_{NI}. This twist orientation is recovered when repeating the cycle of the nematic-isotropic transition. This result indicates that the elastomers can retain the 90° twist mesogen orientation during cross-linking, even after the chiral dopant is removed. The off-square shape of the specimen shown in Figure 2 is a result of thermal deformation, which is described below.

Figure 3 shows the optical micrographs of thermal deformation driven by a temperature change across T_{NI}. When the specimen is heated across T_{NI}, the almost square shape of the cross section normal to the thickness direction (at 40 °C) changes to a diamond-like shape (at 130 °C). Cooling across T_{NI} induces a reverse change in shape. It should be noted that each side of the cross section shown in Figure 3a is almost parallel to the direction in which the bounding mesogens are oriented (which corresponds to the rubbing direction) at the top or bottom surface of the films. The shape of the specimen at 40 °C was almost similar to the shape of the specimen cut out from the sample sheet at 25 °C because the glass transition temperature is around 35 °C. On cooling (heating), the elastomer appears uniaxially stretched (compressed) along the diagonal *a*. We characterize the thermal deformation on the basis of the variations in the length along the diagonals *a* and *b*, which are shown

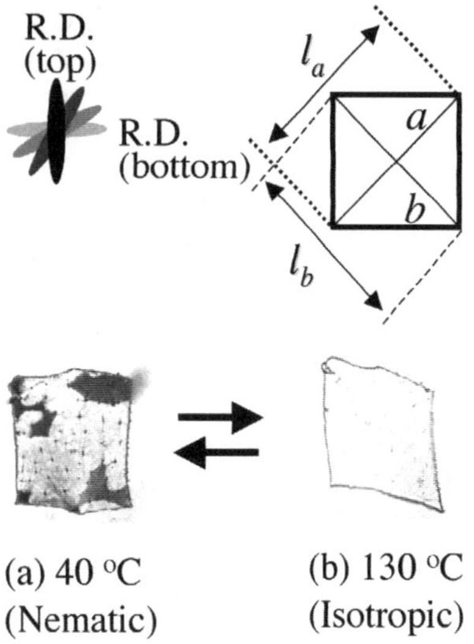

Figure 3. Optical micrographs of the deformation induced by temperature variation between (a) 40 °C and (b) 130 °C across the nematic-isotropic transition temperature (T_{NI}). The specimen was observed under natural light. The diagonals *a* and *b*, and the rubbing directions (R.D.) are indicated in the figure.

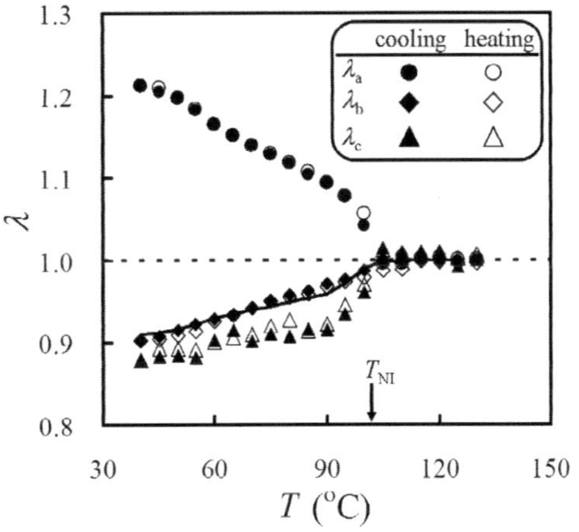

Figure 4. Temperature (T) dependence of the dimensional ratios λ_i (i = a, b, c). The nematic-isotropic transition temperature (T_{NI}) is 103 °C. The solid line indicates the calculated curves of λ_b ($=\lambda_c$) assuming uniaxial stretching of $\lambda_b = \lambda_c = \lambda_a^{-1/2}$.

in Figure 3. The dimensional ratios λ_a and λ_b, whose *reference state is taken in the high-temperature isotropic state* [21], can be defined as follows: $\lambda_i = l_i/l_i^0$ (i = a, b), where l_i is the dimension in the state of interest and the superscript 0 denotes the isotropic state at 130 °C. The corresponding dimensional ratio in the thickness direction (λ_c) was calculated using the formula $\lambda_c = Z_{ab}^0/Z_{ab}$, where Z_{ab} is the cross-sectional area of the *a-b* plane evaluated from video analysis; in this case, the volume of the rubbery specimen was assumed to be constant.

Figure 4 shows the temperature dependence of λ_i (i = a, b, c). No appreciable difference in λ_i between the cooling and heating processes is observed. In the nematic state, the specimen undergoes anisotropic deformation ($\lambda_i \neq 1$). As the temperature decreases, λ_a increases while λ_b and λ_c decrease. The solid lines in Figure 4 represent the values of λ_b ($=\lambda_c$) calculated from the experimental values of λ_a, assuming simple uniaxial stretching in the *a*-direction ($\lambda_b = \lambda_c = \lambda_a^{-1/2}$), i.e., the intermediate direction between the directors of the top and bottom surfaces. The induced deformation is roughly approximated by uniaxial stretching, although a finite difference between the experimental values of λ_b and λ_c indicates that the real deformation is more complicated. The diagonal *a* corresponds to the direction of the summation of the local directors (for the constituent cross sections normal to the helical axis), which gradually rotate by 90° between the top and bottom surfaces. Each cross section is subjected to uniaxial stretching along the local director, but each cross section cannot deform independently of the distortion of the other cross sections. The constituent cross sections in TNEs are strongly bounded by rubber elasticity; this, however, is not the case with low molecular mass LCs or uncrosslinked LC polymers with a similar twist orientation. Therefore, elastomers are expected to exhibit maximum elongation along the direction corresponding to the summation of the local directors.

The inset of Figure 5 shows the T dependence of the absorbance A for the stretching vibration of the terminal cyano group. The absorbance in the figure is reduced by λ_c at each T to consider the change in the thickness with variations in T. In the nematic state ($T < T_{NI}$), a reduction in T results in an increase in A/λ_c. An increase in A/λ_c leads to a decrease in the tilt angle (θ) of the mesogen with respect to the *a-b* plane, because A/λ_c is proportional to $\langle\cos^2\theta\rangle$ as $A/\lambda_c = p\langle\cos^2\theta\rangle$, where p is a constant and the bracket denotes the average over the specimen. Each constituent cross section (normal to the helical axis and parallel to the *a-b* plane) of the twist nematic texture is simply expected to have a uniaxial mesogen alignment with almost the same degree of orientation, whereas the local directors gradually rotate between the top and bottom layers. The orientational order parameter for each cross section is expressed by $S_{local} = [3\langle\cos^2\psi\rangle' - 1]/2$, where ψ is the angle between the

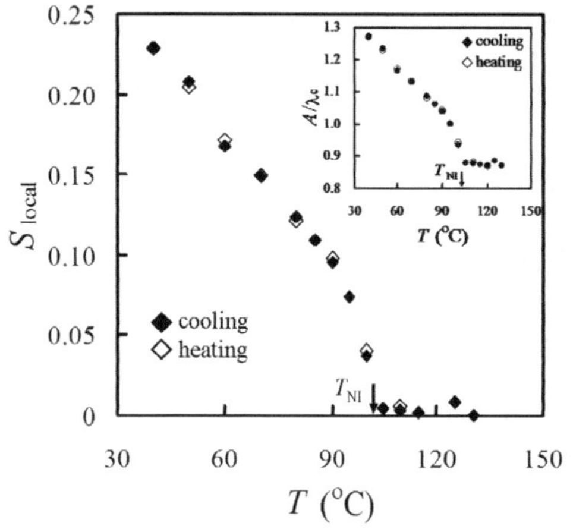

Figure 5. The local order parameter (S_{local}) for the constituent cross sections normal to the helical axis as a function of temperature (T). The inset shows the T dependence of the reduced absorbance (A/λ_c) of the cyano groups at the end of the mesogen. The nematic-isotropic transition temperature (T_{NI}) is 103 °C.

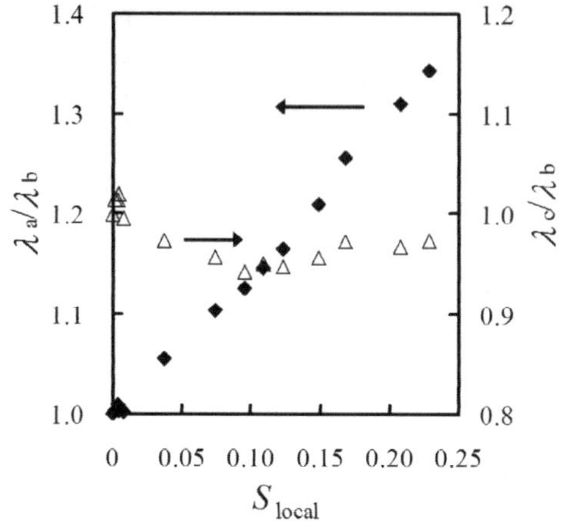

Figure 6. The dimensional parameters λ_a/λ_b and λ_c/λ_b as a function of the local order parameter (S_{local}) for the constituent cross sections normal to the helical axis.

mesogen orientation and the local director, and <....>' represents the average for each cross section. The quantity $<\cos^2\psi>'$ may be approximated by $<\cos^2\theta>$ because of the symmetry of uniaxial orientation, and then S_{local} is correlated with A as $S_{local} = [3A/(p\lambda_c) - 1/2]$. The constant p is determined by the value of A, and $S_{local} = 0$ in the isotropic state with random orientation.

As observed in Figure 5, S_{local} in the nematic state increases with decreasing T. In Figure 6, the ratios λ_a/λ_b and λ_c/λ_b are given as functions of S_{local}. It can be observed that a growth in S_{local} increases the anisotropy for the deformation in the a–b plane, and that the ratio λ_a/λ_b is almost linearly proportional to S_{local}. In contrast, an increase in S_{local} induces no appreciable anisotropy for the deformation in the b–c plane.

A TNE with a desired twist angle (for example, 180° and 270°) can be prepared by the chiral imprinting method. The thermal deformation behaviors of TNEs with different twist angles will provide a further basis to explain the types of deformation driven by the twist orientations. The findings of the deformation induced by the twist orientation and the dependence of the deformation on S_{local} can be applied to the theoretical study of liquid crystal elastomers.

4. Summary

We have shown that a 90° twist nematic elastomer (TNE) prepared by chiral imprinting exhibits macroscopic deformation when the temperature is varied across the nematic–isotropic transition temperature (T_{NI}). The macroscopic deformation reflects the change in mesogen alignment between the twist nematic and random orientations. The macroscopic deformation is roughly approximated by a uniaxial deformation in a direction at an angle of 45° relative to the mesogen orientations at the top and bottom surfaces. The distortion increases with a reduction in temperature, i.e., an increase in orientational order.

Acknowledgments

The authors would like to thank Merck Ltd., Japan, for providing the chiral dopant. This work was partly supported by a Grant-in-Aid on Priority Area "Soft Matter Physics" (No. 19031014) and a

Grant-in-Aid for Scientific Research (B) (No. 16750186) from the Ministry of Education, Culture, Sports, Science and Technology (MEXT) of Japan. This research was also supported in part by the Global COE Program "International Center for Integrated Research and Advanced Education in Materials Science" (No. B-09) of MEXT of Japan, administrated by the Japan Society for the Promotion of Science.

References and Notes

[1] Warner M, Terentjev EM 2007 *Liquid Crystal Elastomers Revised Edition*. (London, Clarendon Press)
[2] Urayama K 2007 *Macromolecules* **40** 2277
[3] Kupfer J, Finkelmann H 1991 *Makromol. Chem. Rapid Commun.* **12** 717
[4] Wermter H, Finkelmann H 2001 *e-Polymers* 013
[5] Clarke SM, Hotta A, Tajakhsh AR, Terentjev EM 2001 *Phys. Rev. E* **64** 061702
[6] Thomsen DL, Keller P, Naciri J, Pink R, Jeon H, Shenoy D, Ratna BR 2001 *Macromolecules* **34** 5868
[7] Hiraoka K, Sagano W, Nose T, Finkelmann H 2005 *Macromolecules* **38** 7352
[8] Oswald P, Pieranski P *Nematic and Cholesteric Liquid Crystals*. CRC Press: Boca Raton, 2005.
[9] Spillmann CA, Naciri J, Martin BD, Farahat W, Herr H, Ratna BR 2007 *Sensors and Actuators A* **133** 500
[10] de Gennes PG 1969 *Phys. Lett. A* **28** 765
[11] Tsutsui T, Tanaka R, Tanaka J 1979 *J. Polym. Sci., Polym. Lett. Ed.* **17**, 511.
[12] Mol GN, Harris KD, Bastiannsen CWM, Broer DJ 2005 *Adv. Funct. Mater.* **15** 1155
[13] Hasson CD, Davis FJ, Mitchell GR 1998 *Chem. Commun.* 2515
[14] Courty S, Tajbakhsh AR, Terentjev EM 2003 *Phys. Rev. Lett.* **91** 085503
[15] Marty JD, Gornitzka H, Mauzac M 2005 *Eur. Phys. J. E* **17** 515
[16] Courty S, Tajbakhsh AR, Terentjev EM 2006 *Phys. Rev. E* **73** 011803
[17] Mao Y, Warner M 2000 *Phys. Rev. Lett.* **84** 5335
[18] Shibaev VP, Kostromin SG, Plate NA. 1982 *Eur. Polym. J.* **18** 651
[19] Grandjean F. 1921 *C. R. Hebd. Seances Acad. Sci.* **172** 71
[20] Cano R, Chatelain P 1971 *C. R. Hebd. Seances Acad. Sci.* **253B** 1815
[21] Specimens with a square cross section *in the high-temperature isotropic state* are more suitable for straightforward demonstration of thermal deformation. However, we were unable to prepare such specimens due to the difficulty involved in cutting the samples at high temperatures above 100 °C.

Journal of Physics: Conference Series **184** (2009) 012023 doi:10.1088/1742-6596/184/1/012023

Density functional simulations for self-assembling micellar structures in amphiphilic block copolymer solutions

Takashi Uneyama

JST-CREST, Institute for Chemical Research, Kyoto University, Uji 611-0011, Japan

E-mail: uneyama@scl.kyoto-u.ac.jp

Abstract. Density functional simulations are performed to study various self-assembling micellar structures in amphiphilic block copolymer solutions. It was shown that onion structures (stacked multilayer structures) are formed when the hydrophilic interaction is weak. Simulation results imply that the onion structures are different from monolayer vesicle structures which is formed when the hydrophilic interaction is rather strong. The difference and similarity between the vesicle structures and onion structures, or between their kinetic formation pathways are discussed.

1. Introduction

It is known that block copolymers form various self-assembling micellar structures when dissolved in solvents [1]. For example, well-known micellar structures such as spheres, cylinders, or vesicles (closed bilayer structures) are formed spontaneously in amphiphilic block copolymer solutions. However, it is reported that more complicated micellar structures [2, 3] can be also formed by block copolymer solutions.

The morphologies of such self-assembling structures depend on various parameters such as the volume fraction of the polymers, the block ratio, or the solvent quality (strength of the interaction). Although many experiments [1] and simulations [4, 5] have been performed so far, the effects of these parameters on the resulting self-assembling structures are still not well understood.

In this work, the density functional simulations for amphiphilic block copolymer solutions were performed to study the effects of the interaction parameters. Special attention was payed on the effect of the strength of the solvophilic interaction. The difference and similarity between the onion structures (multilayer vesicle-like structures) and the single layer vesicle structures are then discussed. It is shown that they are qualitatively different while they share several common properties.

2. Model

We consider blends of amphiphilic AB diblock copolymers (which consists of the solvophilic A-subchain and the solvophobic B-subchain) and S solvents (the solvents are small molecules or polymers which have the selective interaction) as models of amphiphilic block copolymer solutions.

In this work, the density functional model for amphiphilic block copolymer solutions [6] is employed. In the density functional model, the state of the amphiphilic block copolymer solution is expressed by using three density fields $\rho_i(\boldsymbol{r})$ (i = A, B, S). The free energy of the system can be described as a functional of the density fields. The model free energy functional can be expressed as follows.

$$\begin{aligned}\frac{\mathcal{F}[\{\rho_i\}]}{k_B T} &= \sum_{i,j\,(=\mathrm{A,B})} \int d\boldsymbol{r}d\boldsymbol{r}'\, 2\sqrt{f_i f_j} A_{ij} \tilde{\mathcal{G}}(\boldsymbol{r}-\boldsymbol{r}') \sqrt{\rho_i(\boldsymbol{r})\rho_j(\boldsymbol{r}')} \\ &+ \sum_{i\,(=\mathrm{A,B})} \int d\boldsymbol{r}\, f_i C_{ii} \rho_i(\boldsymbol{r}) \ln \rho_i(\boldsymbol{r}) + \int d\boldsymbol{r}\, C_{\mathrm{SS}} \rho_{\mathrm{S}}(\boldsymbol{r}) \ln \rho_{\mathrm{S}}(\boldsymbol{r}) \\ &+ \int d\boldsymbol{r}\, 4\sqrt{f_{\mathrm{A}} f_{\mathrm{B}}} C_{\mathrm{AB}} \sqrt{\rho_{\mathrm{A}}(\boldsymbol{r})\rho_{\mathrm{B}}(\boldsymbol{r})} + \sum_{i\,(=\mathrm{A,B,S})} \int d\boldsymbol{r}\, \frac{b^2}{6} \left| \nabla \sqrt{\rho_i(\boldsymbol{r})} \right|^2 \\ &+ \sum_{i,j\,(=\mathrm{A,B,S})} \int d\boldsymbol{r}\, \frac{\chi_{ij}}{2} \rho_i(\boldsymbol{r})\rho_j(\boldsymbol{r}) + \int d\boldsymbol{r} \frac{P(\boldsymbol{r})}{2} \left[\rho_{\mathrm{A}}(\boldsymbol{r}) + \rho_{\mathrm{B}}(\boldsymbol{r}) + \rho_{\mathrm{S}}(\boldsymbol{r}) - 1\right]\end{aligned} \tag{1}$$

where b is the segment size, f_i is the block ratio of the i-subchain, A_{ij} and C_{ij} are constants determined from b and f_i. $\tilde{\mathcal{G}}(\boldsymbol{r})$ is the nonlocal entropic interaction kernel which represents the effect of the connectivity between two blocks. χ_{ij} is the Flory-Huggins χ parameter between the i- and j-species, and $P(\boldsymbol{r})$ is the Lagrangian multiplier field which is determined to satisfy the incompressible condition $\rho_{\mathrm{A}}(\boldsymbol{r}) + \rho_{\mathrm{B}}(\boldsymbol{r}) + \rho_{\mathrm{S}}(\boldsymbol{r}) = 1$. For the time evolution, the following simple dynamic density functional equation [7] is assumed.

$$\frac{\partial \rho_i(\boldsymbol{r},t)}{\partial t} = \nabla \cdot \left[\frac{\rho_i(\boldsymbol{r})}{\zeta} \nabla \frac{\delta \mathcal{F}[\{\rho_i\}]}{\delta \rho_i(\boldsymbol{r})} \right] + \xi_i(\boldsymbol{r},t) \tag{2}$$

where ζ is the friction coefficient for a monomer (which is assumed to be the same for all the monomer species) and $\xi_i(\boldsymbol{r},t)$ is the thermal noise which satisfies the following fluctuation-dissipation type relation.

$$\langle \xi_i(\boldsymbol{r},t) \rangle = 0, \quad \langle \xi_i(\boldsymbol{r},t)\xi_j(\boldsymbol{r}',t') \rangle = -2\tilde{\beta}^{-1} k_B T \delta_{ij} \nabla \cdot \left[\frac{\rho_i(\boldsymbol{r})}{\zeta} \nabla \delta(\boldsymbol{r}-\boldsymbol{r}') \right] \delta(t-t') \tag{3}$$

Here, $\tilde{\beta}^{-1} \leq 1$ is the factor which determines the noise amplitude (the effective temperature of the noise reduces to $\tilde{\beta}^{-1}T$).

3. Simulations

So far, the density functional simulations for amphiphilic block copolymer systems with sufficiently strong solvophilic systems are performed [6,7]. For sufficiently solvophilic cases, it was shown that several well-known micellar structures can be actually formed and the morphologies depend on several parameters such as the volume fraction, the block ratio, or the interaction strength.

In this work, to consider the effect of hydrophilic interaction in more detail, simulations for not strong solvophilic interaction cases are performed. Thermodynamically stable structures can be directly obtained by minimizing the free energy (1) numerically (statics simulations). Dynamics simulations can be performed by solving the discretized version of eq (2) together with eq (1). It is empirically known that the resulting structures are qualitatively the same both for the static and dynamic simulations.

Snapshots for a typical dynamic simulation for a rather solvophobic interaction case are shown in Figure 1. The parameters for the simulation are as follows: the volume fractions

$\bar{\rho}_{\mathrm{AB}} = 0.2, \bar{\rho}_{\mathrm{S}} = 0.8$, the degrees of polymerization $N_{\mathrm{AB}} = 10, N_{\mathrm{S}} = 10$, the block ratios $f_{\mathrm{A}} = f_{\mathrm{B}} = 0.5$, and the χ parameters $\chi_{\mathrm{AB}} = 1.75, \chi_{\mathrm{AS}} = 0.5, \chi_{\mathrm{BS}} = 2.5$, the friction coefficient $\zeta = 1$, the temperature $k_B T = 1$, and the noise parameter $\tilde{\beta}^{-1} = 0.04$. The initial condition (at $t = 0$) is the homogeneous state and the system size is $24 \times 24 \times 24$ (periodic boundary condition). The time step size is $\Delta t = 0.0025$ and the lattice size is $\Delta x = 0.5$. The simulation is performed from $t = 0$ to $t = 3125$.

From Figure 1, it can be observed that small micellar structures are formed rapidly from the homogeneous initial state (Figure 1 (a),(b)), and the structures slowly grow into multilayer onion structures mainly by collision of small structures (Figure 1 (c),(d)). The observed structures look similar to ones obtained by the self consistent field simulations for diblock copolymer solutions [8].

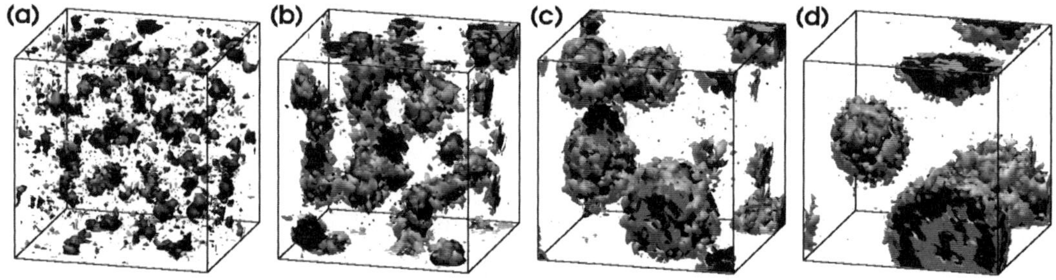

Figure 1. Snapshots of the dynamic density functional simulation for the onion structure formation. Grey and black surfaces mean the isodensity surfaces for solvophilic and solvophobic blocks ($\rho_{\mathrm{A}}(\boldsymbol{r}) = 0.5$ and $\rho_{\mathrm{B}}(\boldsymbol{r}) = 0.5$) , respectively. (Solvents are not explicitly shown.) The time for each snapshot corresponds to (a) $t = 6.25$, (b) $t = 62.5$, (c) $t = 625$, and (d) $t = 3125$.

4. Discussions

From the previous simulation results [6, 7], we find that spherical micelles, cylindrical micelles, and bilayer vesicles are formed when the solvophilic interaction is strong. The simulation results by this work implies that onions (multilayer vesicle-like structures) are formed when the solvophilic interaction is weak or the block copolymer is solvophobic. Also we find that the formation pathway of the onion structures is different from one of the vesicle structures. In the vesicle formation case, first, small micellar structures are formed from the homogeneous initial state. Then small structures grow into micellar bilayer structures by collisions, and finally bilayer structures are closed to form vesicles. This is the conventional vesicle formation process [9] and is also observed in particle simulations [4, 10].

In the onion formation case, the early stage is almost the same as the vesicle formation case. Small micellar structures are formed from the homogeneous state. The small structures grow into larger droplet like structures by collisions. Layer structures are rearranged in the droplets and thus multilayer structures are formed. Onion structures become larger by further collisions and rearrangements. Also, structures can grow by the evaporation-condensation (or the Ostwald ripening) type process. (Both the collision-rearrangement and evapolation-condensation growth processes are slow.) It will be reasonable to consider that the onion structures are microphase separation structures in block copolymer rich droplets [2] rather than usual micellar structures. This will be a natural interpretation because the micellar structures (or the block copolymer rich domains) always have spherical droplet like shapes, and there is rather strong interfacial energy between the micellar droplet phase and the outer solvent phase. In fact, we obtain packed cylinder like structures in droplet like regions if we change the block ratio. These results are consistent with the simulation results by other continuum field models [8, 11].

Therefore, the two formation pathways are different in the late stage while they are quite similar in the early stage. These formation pathways are schematically shown in Figure 2. While the pathways are different, there are still similarity between two cases even in the late stage. Each structure has closed spherical shell like structures and they are formed from disk like layer structures. The layer closure processes look quite similar for two cases. The layers have bending energy, and the closure processes are driven by other thermodynamic forces. For the vesicle formation case, the driving force is the line tension energy at the bilayer edges [9]. However, for the onion formation case, it will be reasonable to consider that the driving force is the surface energy of the micellar droplets because the shapes of droplets are always kept to be spherical and the contribution of the line tension energy is generally not large for relatively small structures. Thus the driving forces of the structure formations are different for two processes. This is consistent with the fact that the characteristic size of monolayer onion structures is smaller than one of the vesicles because the contribution of the surface energy is strong even for small structures, compared with the line tension energy.

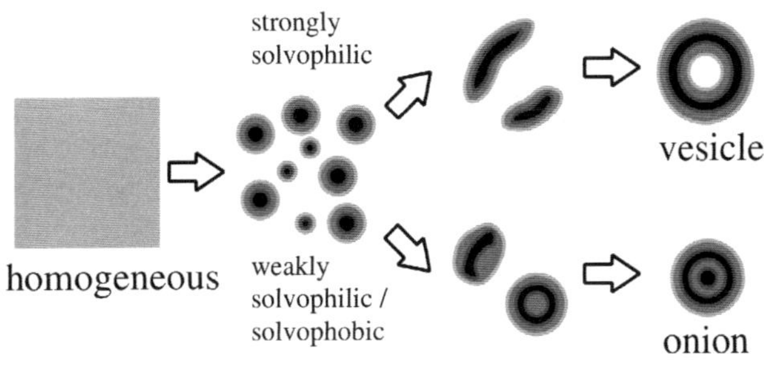

Figure 2. Schematic draw of the formation pathways of a monolayer vesicle and a multilayer onion. Grey and black colors mean the solvophilic and solvophobic subchains, respectively. The upper pathway corresponds to the strongly solvophilic systems whereas the lower pathway corresponds to the weakly solvophilic or solvophobic systems.

5. Conclusion

The density functional simulations for amphiphilic block copolymer solutions were performed with various solvophilic interaction strengths. It was shown that when the solvophilic interaction is weak or the block copolymer is solvophobic, stacked multilayer onion structures are formed instead of well-known micellar structures such as vesicles. We found that the formation pathway of the onion structures is different from one of the vesicles, and discussed the difference and similarity between them. To make the difference and similarity clearer we will need to perform simulations with various interaction parameters and discuss thermodynamic properties in more detail. Further investigation will be the subject of future work.

References

[1] Hamley I W 2005 *Block Copolymers in Solution: Fundamentals and Applications* (Chichester: Wiley)
[2] Fraaije J G E M and Sevink G J A 2003 *Macromolecules* **36** 7891–7893
[3] Jain S, Gong X, Scriven L E and Bates F S 2006 *Phys. Rev. Lett.* **96** 138304
[4] Yamamoto S, Maruyama Y and Hyodo S 2002 *J. Chem. Phys.* **116** 5842–5849
[5] He X, Liang H, Huang L and Pan C 2004 *J. Phys. Chem. B* **108** 1731–1735
[6] Uneyama T and Doi M 2005 *Macromolecules* **38** 5817–5825
[7] Uneyama T 2007 *J. Chem. Phys.* **126** 114902
[8] Sevink G J A and Zvelindovsky A V 2005 *Macromolecules* **38** 7502–7513
[9] Fromherz P 1983 *Chem. Phys. Lett.* **94** 259–266
[10] Noguchi H and Takasu M 2001 *Phys. Rev. E* **64** 041913
[11] Ohta T and Ito A 1995 *Phys. Rev. E* **52** 5250–5260

Graftpolyrotaxane: Graft Polymer Possessing Movable Graft Chains on Cyclodextrins as The Polyrotaxane Wheels

T Takashima, K Hinoue, N Kihara, M Hayashi, Y Koyama, and T Takata[1]
Department of Organic and Polymeric Materials, Tokyo Institute of Technology, Ookayama, Meguro, Tokyo 152-8552, Japan

E-mail: ttakata@polymer.titech.ac.jp

Abstract. Graftpolyrotaxanes having graft chains on the wheel components were prepared from main chain-type polyrotaxane having monofunctional cyclodextrin wheels and terminal end-reactive poly(ethylene glycol) (PEG). The well-defined polyrotaxane having mono(6-hydroxyl) permethyl-α-CD (monoOH-α-CD) as the wheel components and a polytetrahydrofuran (PTHF) axle with bulky end-capping groups was synthesized by the two pathways; (i) synthesis by solid-state end-capping of pseudopolyrotaxane consisting of monoOAc-α-CD and hydroxy-terminated PTHF and (ii) one-pot synthesis using monoOH-α-CD and amine-terminated PTHF in water. The grafting reaction on the wheel moieties of the resulting polyrotaxane was carried out by acylation reaction with PEG-mono-carboxylic acid derivatives in high conversion yields. To estimate the relative rotary movement of CDs and the axle polymer in the polyrotaxane and graftpolyrotaxane, the rotational correlation time (τ_c) of polyrotaxanes was estimated from the longitudinal relaxation time (T_1) of ^{13}C NMR. The results clearly indicate that the graftpolyrotaxane described here is a new type of graft polymer in terms of the special circumrotation behavior such as the independent rotary movement of CDs with the graft chain and the axle polymer.

1. Introduction

Supramolecules with interlocked structures have attracted great interests experimentally and theoretically [1]. One of the typical examples can be provided by cyclodextrin (CD)-containing main-chain type polyrotaxane, in which many cyclic molecules are threaded on a single polymer chain and are trapped by capping the chain end with bulky groups. Networked polymer of polyrotaxane obtained by cross-linking of wheel components on the main-chain type polyrotaxane has unique characteristics based on the interlocked structures of the cross-link points, such as highly swelling nature for several solvents, high elasticity, and high stress-releasing ability [2]. Graft polyrotaxane, a new type of polymer characterized by the movable (translative and cicumrotative) graft chains, is an intriguing class of polyrotaxanes, which can be provided by a grafting protocol to wheel components of the main chain-type polyrotaxane (Figure 1). Although several studies of graftpolyrotaxane directed toward drug delivery systems using a biodegradable property [3] or high stress releasing materials exploiting the sliding property of wheel components [4] have been extensively studied, the dynamic characteristic of graftpolyrotaxane has never been studied owing to the high structural disorderness of graftpolyrotaxane [5]. Namely, the grafting reaction onto CDs of polyrotaxane naturally involves non-regioselective functionalization, since CD has many free hydroxyl groups.

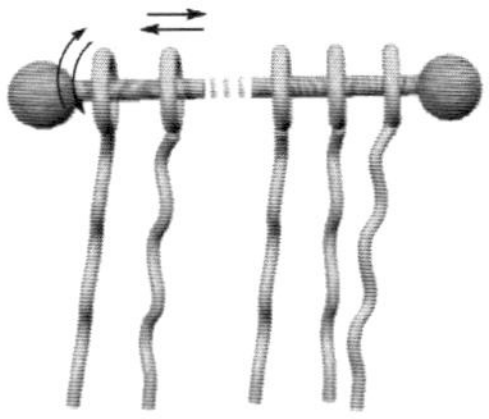

Figure 1. Structure of graftpolyrotaxane.

Recognizing these issues, we planned the synthesis of structure-well-defined graftpolyrotaxane by exploiting monoOH-α-CD as a crucial component of main chain-type polyrotaxane. Herein, we wish to report the synthesis, characterization, and dynamic property of graftpolyrotaxane.

2. Experimental Section

2.1. Materials and Methods. MonoOH-α-CD **5** and polytetrahydrofuran bisamine terminated (PTHFBA, M_n = 9400, M_w/M_n = 1.1) were prepared according to the literature [6,7]. Dichloromethane was distilled from CaH_2 and stored over MS 4A. Commercially available dehydrated solvents, tetrahydrofuran (THF), and dimethylformamide (DMF) were used without further purification. All other reagents from commercial sources were used as obtained. ^{1}H (270, 300 and 400 MHz) and ^{13}C (75 and 100 MHz) NMR spectra were recorded on JEOL JMN 270, BRUKER-SPECTROSPIN 300, and JEOL AL-400 spectrometers using $CDCl_3$ as the solvent and tetramethylsilane as the internal standard. Molecular weight and its distribution were measured by a gel permeation chromatography (GPC) on a JASCO Gulliver system equipped with two consecutive linear polystyrene gel columns (TOSOH TSKgel G2000HXL and GMHXL) at 30 °C, eluted with $CHCl_3$ at a flow rate of 0.85 mL/min, and calibrated using polystyrene standards. IR spectra were recorded on a JASCO FT/IR-230 spectrometer. Thermogravimetry (TG) was performed on a Shimadzu TGA-50 instrument at a heating rate of 10 °C/min under N_2 atmosphere (flow rate of 50 mL/min) to determine decomposition temperature T_{d5} at which 5% weight loss was observed. Differential scanning calorimetry (DSC) was carried out using Shimadzu DSC-60 instrument at a heating rate of 10 °C/min under N_2 atmosphere (flow rate of 50 mL/min) to determine glass transition temperature T_g.

Preparation of Mono(6-acetoxy)-permethyl-α-CD (monoOAc-α-CD, 1): To a solution of MonoOH-α-CD **5** (1.9 g, 1.6 mmol) in THF (10 mL) was added Ac_2O (605 μL, 6.4 mmol) and pyridine (516 μL, 6.4 mmol) at room temperature. The mixture was stirred for 12 h at the same temperature. The reaction was stopped by the addition of sat. aq. $CuSO_4$ (200 mL) and the layers were separated. The products were extracted with $CHCl_3$ (100 mL x 2). The combined organic layer was washed with brine, dried over $MgSO_4$, and concentrated in vacuo to give MonoOAc-α-CD (**1**, 1.7 g) in 85% yield: m.p. 240–247 °C, ^{1}H NMR (270 MHz, $CDCl_3$) δ 5.06 (d, J = 3.5 Hz, 6H), 3.90–3.16 (m, 87H), 2.07 (s, 3H), IR (neat): 1742, 1237 cm^{-1}.

Synthesis of Polyrotaxane by Solid-State End-Capping Reaction: MonoOAc-α-CD (**1**, 1.5 g, 1.2 mmol) and polytetrahydrofuran (PTHF) (M_n 2000, 75 mg, 0.038 mmol) were suspended in water and sonicated for 1 h at room temperature and allowed to stand overnight at the same temperature. The resulting mixture was separated by centrifugation, washed with H_2O, and dried under vacuum to give pseudopolyrotaxane (**2**, 726 mg, 100%, on the basis of PTHF). The crude material was used for next reaction without further purification: ^{1}H NMR (270 MHz, acetone-d$_6$) δ 5.03 (brd, C(1)H of CD moieties), 3.86–3.03 (m, C(2–6)H and MeO of CD moieties, and PTHF), 2.18 (s, AcO), 1.61 (m, PTHF).

A mixture of the crude pseudopolyrotaxane (**2**, 600 mg, 0.019 mmol), tritylphenyl isocyanate (205 mg, 0.57 mmol) [8], and dibutyltindilaulate (DBTDL, 1.1 mL, 1.9 mmol) was well ground in an agate mortar at room temperature. After 30 min grinding, the resulting mixture was dissolved in $CHCl_3$ and filtrated. The filtrate was precipitated into Et_2O to give a white solid which was dissolved in $CHCl_3$ and precipitated into MeOH to give the corresponding polyrotaxane as a white solid (**3**, 107 mg) in 43% yield: ^{1}H NMR (270 MHz, $CDCl_3$ in the presence of a catalytic amount of TFA) δ 7.37–7.06 (m, ArH of end-capping group), 5.01 (brd, C(1)H of CD moieties), 3.98–3.01 (m, C(2–6)H and MeO of CD moieties, and PTHF), 2.18 (s, AcO), 1.61 (brd, PTHF).

*Synthesis of Reactive Polyrotaxane Consisting of MonoOH-α-CD (**4**):* To a solution of the resulting polyrotaxane (**3**, 90 mg, 6.9 mmol) in $CHCl_3$ (2.0 mL) was added a solution of saturated K_2CO_3 in MeOH (2.0 mL) at room temperature. The mixture was warmed to 50 °C, stirred overnight, and concentrated in vacuo. The crude materials dissolved in $CHCl_3$ was filtrated through Celite® pad. The filtrate was concentrated in vacuo to give the reactive polyrotaxane with monoOH-α-CD (**4**, 88 mg) in a quantitative yield: ^{1}H NMR (270 MHz, $CDCl_3$ in the presence of a catalytic amount of TFA) δ 7.42–7.07 (m, ArH of end-capping group), 5.03 (brd, C(1)H of CD moieties), 3.87–2.94 (m, C(2–6)H and MeO of CD moieties, and PTHF), 1.60 (brd, PTHF).

Preparation of Reactive Polyrotaxane Consisting of MonoOH-α-CD (7) by One-pot Synthesis: MonoOH-α-CD (**5**, 1.2 g, 1.0 mmol) and PTHFBA (140 mg, 2.0 mmol) were suspended in water and sonicated for 30 min at room temperature. To the reacting mixture containing colorless precipitate was directly added 3,5-dimethylphenyl isocyanate (110 µl, 0.75 mmol), and the mixture was stirred for 1 h at room temperature. Colorless precipitate was separated by centrifugation and dried under vacuum. Crude product was purified by a gel permeation chromatography (GPC) and precipitation with Et_2O to yield pure polyrotaxane (**7**, 108 mg, M_n = 28000, M_w/M_n = 1.3). Coverage ratio of polyrotaxane was approximately 44% estimated by comparison of the integration of C(1)H of monoOH-α-CD and methylene proton of PTHFBA: ^{1}H NMR (400 MHz, $CDCl_3$, 298 K) δ 5.02 (brd, 6H, C(1)H), 4.11–3.13 (m, 123H, C(2-6)H, O(2,3,6)CH_3, CH_2O of PTHF), 1.60 (m, 13H, CH_2 of PTHF) ppm, ^{13}C NMR (100 MHz, $CDCl_3$, 298 K) δ 100.3 (C1), 100.0, 99.9, 99.7, 82.2, 81.9, 81.3 (C2), 71.3, 71.1, 70.5 (PTHF), 61.8, 58.9, 57.8, 57.7, 26.4 (PTHF) ppm, IR (KBr) 3482, 2931, 1459, 1366, 1302, 4037, 1039, 973, 857, 701, 559 cm^{-1}, T_g –55 °C, T_{d5} 355 °C.

*Preparation of polyethylene glycol mono oleylether mono carboxylic acid (OPEGCOOH, **8**):* To a solution of polyethylene glycol mono oleylether (OPEG, 3.5 g, 3.0 mmol) and tetramethylpiperidine *N*-oxide (TEMPO, 0.59 g, 3.8 mmol) in H_2O (20 mL) was added dropwise 5 w% aq. NaOCl (60 mL, 37 mmol) at 0 °C. The reaction mixture was warmed to room temperature and stirred for 15 min. The mixture was diluted with CH_2Cl_2 and quenched by the addition of EtOH. The product was extracted with CH_2Cl_2 (x 3), and the combined organic extracts were washed with brine, dried over $MgSO_4$, and concentrated in vacuo. The crude materials were purified by repeated precipitation from hexane to give **8** (2.50 g, 72%, M_n 1200): ^{1}H NMR (400 MHz, $CDCl_3$, 298 K) δ 5.38–5.33 (m, 1H, CH=CH), 4.16 (s, 2H, OCH_2COO), 3.82–3.43 (m, 84H, OCH_2CH_2O), 2.02–1.99 (m, 4H, CH_2), 1.57 (t, J = 6.8 Hz, 2H, CH_2), 1.41–1.25 (m, 27H, CH_2), 0.88 (t, J = 6.6 Hz, 3H, CH_3) ppm, T_g 22 °C.

*Synthesis of graftpolyrotaxane (10) utilizing OPEGCOOH (**8**):* To a solution of polyrotaxane (**7**, 30 mg, 20 µmol on the basis of CD unit), **8** (220 mg, 200 µmol) and *N,N*-dimethylaminopyridine (DMAP, 25 mg, 200 µmol) in CH_2Cl_2 (1.0 mL) was added *N*-(3-dimethylaminopropyl)-*N'*-ethylcarbodiimide hydrochloride (EDC·HCl, 380 mg, 2.0 mmol) at room temperature. After stirring for 4 d at the same temperature, the mixture was quenched by the addition of H_2O. The product was extracted with CH_2Cl_2 (x 3), and the combined organic extracts were washed with brine, dried over $MgSO_4$, and concentrated in vacuo. The crude materials were purified by a gel permeation chromatography to give the graft polyrotaxane **10** (7.2 mg, 14%, M_n 16000): ^{1}H NMR (400 MHz, $CDCl_3$, 298 K) δ 5.38–5.33 (m, 1H, CH=CH), 5.02 (brd, 6H, C(1)H), 4.53 (s, 1H, C(5)H), 4.39 (brd, 1H, C(5)H), 4.12 (brd, 2H, CH_2COO), 4.11–3.13 (m, 123H, C(2-6)H, O(2,3,6)CH_3, CH_2O of PTHF), 2.02–1.99 (m, 4H, CH_2), 1.61 (brd, 2H, CH_2), 1.41–1.25 (m, 27H, CH_2), 0.88 (t, J = 6.6 Hz, 3H, CH_3) ppm, ^{13}C NMR (100 MHz, $CDCl_3$, 298 K) δ 169.9, 129.8, 100.4 (C1), 100.0, 99.8, 82.4, 82.3, 82.1, 81.2 (C2), 71.5, 71.4, 71.2, 70.8, 70.5 (PTHF), 61.8, 59.0, 57.8, 31.8, 29.7, 29.6, 29.5, 29.3, 27.1, 26.4 (PTHF), 26.0, 22.6,

14.1 ppm, IR (KBr) 3512, 2926, 2861, 1759, 1655, 1560, 1459, 1370, 1301, 1252, 1111, 1041, 977, 954, 857, 703, 564 cm^{-1}, no T_g, T_{d5} 325 °C.

Longitudinal relaxation time measurements: ^{13}C NMR measurements of **7** and **10** were performed to determine the longitudinal relaxation time (T_1) by using resonance frequencies for ^{13}C at w_r = 75 MHz and 100 MHz at 298 K via a conventional inversion recovery pulse sequence, 90°–τ–180°, under the deuterium lock mode. The rotational correlation time (τ_c) for each ^{13}C nucleus in the polyrotaxanes was determined with the obtained T_1 values at different resonance frequencies according to eq. 1:

$$\frac{1}{NT_1} = \left(\frac{\mu_0}{4\pi}\right)^2 \frac{\gamma_H^2 \gamma_C^2 h^2}{40\pi^2 r_{CH}^6} \left(\frac{\tau_c}{1+9(\omega_r \tau_c)^2} + \frac{3\tau_c}{1+(\omega_r \tau_c)^2} + \frac{6\tau_c}{1+25(\omega_r \tau_c)^2} \right) \qquad (1)$$

where N, μ_0, γ_H, γ_C, h, and r_{CH} are the number of ^{1}Hs directly bound to ^{13}C, permeability of free space (4π x 10^{-7} H·m^{-1}), gyromagnetic ratio of ^{1}H (2.675 x 10^8 rad·T^{-1}·s^{-1}), gyromagnetic ratio of ^{13}C (0.688 x 10^8 rad·T^{-1}·s^{-1}), Planck's constant (6.626 x 10^{-34} J·s), and distance between ^{1}H and ^{13}C, respectively, 0.1091 nm for methylene protons of a PTHF and 0.1092 nm for C1-H and C2-H of the α-CDs. These distances were calculated by Jaguar 6.5 (Schrödinger, DFT, 6-31G**, B3LYP).

3. Results and Discussion

3.1. Synthesis of Main Chain-Type Polyrotaxane. Scheme 1 features the synthetic routes of structure-definite main chain-type polyrotaxanes via two pathways. One approach (top) involves the synthesis of polyrotaxane **4** by end-capping of the pseudopolyrotaxane **2** comprising of monoOAc-α-CD (**1**) and PTHF with electrophiles in the solid–state [9]. Protection of the wheel hydroxyl group by acetylation to **1** (85% yield) was followed by the complexation with PTHF (M_n 2000) to produce pseudopolyrotaxane (**2**, 100% yield). After considerable experimental works for the end-capping reactions of **2**, we turned out that the homogeneous systems in organic solvents such as DMAc were unfruitful, *i.e.*, dethreading of the wheel components occurs at a rate much faster than the end-capping at the axle termini to afford the end-reacted PTHF without any CD moiety. We thereby subjected it to solid-state reaction. A mixture of **2**, bulky isocyanate (R–NCO), and DBTDL was well ground in an agate mortar at room temperature. After 30 min grinding, polyrotaxane **3** was isolated in 43% yield as the Et_2O- and MeOH-insoluble part. The ^{1}H NMR spectrum of **3** suggests that the two terminal hydroxyl groups were completely converted to urethane groups and that the polymer had 8.2 wheels per axle on average (degree of coverage of the axle: θ 56%). Although the grinding was prolonged to 90 min, neither θ nor yield increased. The alkaline methanolysis of **3** quantitatively yielded the structure-definite reactive polyrotaxane **4** with the hydroxyl-functionalized wheels.

The other synthesis of polyrotaxane (bottom) utilizes one-pot protocol exploiting our recent approach [10], in which monoOH-α-CD **5** and PTHFBA were used in H_2O. A mixture of **5** and PTHFBA (M_n 9400, M_w/M_n 1.1) in H_2O was subjected to the sonication for 30 min at room temperature to give pseudopolyrotaxane **6** as white precipitates, which was followed by the direct addition of 3,5-dimethylphenylisocyanate (5.0 equiv.) and stirring for 1 h at room temperature, yielding polyrotaxane **7** in 16% yield (M_n 28000, M_w/M_n 1.3, estimated by GPC based on polystyrene standards, θ 44%). The structure was confirmed by ^{1}H NMR spectrum.

3.2. Synthesis of Graft Polyrotaxane. Scheme 2 shows the synthesis of graft polyrotaxane **10**. Several classical esterifications of **7** via graft-onto protocol were examined (Table 1). After many experiments, we found that this goal was achieved by exploiting OPEGCOOH **8** (M_n 1300, M_w/M_n 1.2) with excess amount of EDC·HCl as dehydrating agent in the presence of DMAP to allow the effective introduction of graft chains on the wheel (entry 1). The conversion yield of **10** was determined from the integral ratio of ^{1}H NMR spectrum. When the grafting reaction was carried out at 40 °C (entry 2), the molecular weight distribution of **10** became broad probably owing to the occurrence of the partial cross-linking reaction at the olefin moieties. On the other hand, treatment of **7** bearing higher coverage ratio (θ 72%) afforded the corresponding graftpolyrotaxane (**10**) in a low grafting conversion yield due mostly like to the coverage of the reacting points (entry 3).

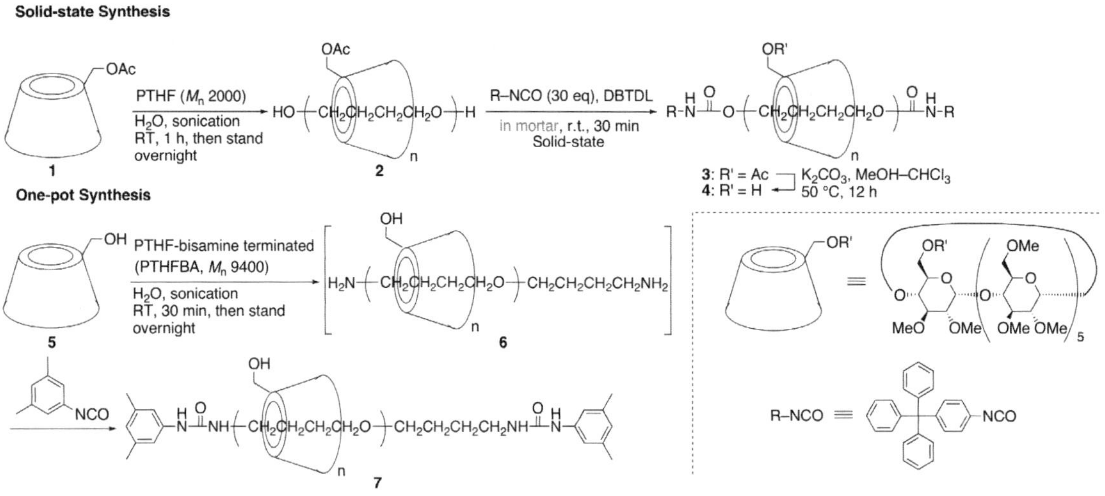

Scheme 1. Two synthetic approaches to structure-definite polyrotaxanes (**4** and **7**).

Although OPEGCOCl **9** was also used to give **10** in a high conversion yield, the reaction needed higher temperature (40 °C) rather than the case of **8** (entry 5). The graft polyrotaxane **10** showed good solubility in various solvents such as THF, DMF, acetone, $CHCl_3$ and CH_2Cl_2 in contrast to polyrotaxane **7** which was soluble only in $CHCl_3$ and CH_2Cl_2.

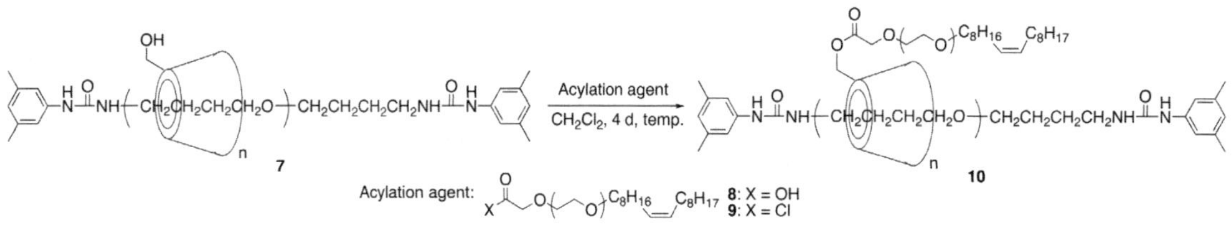

Scheme 2. Synthesis of graftpolyrotaxane **10** according to the grafting-onto protocol.

Table 1. Synthesis of graftpolyrotaxane **10** by grafting of **7**.

Entry	Graft Chain	Reagents	Temp. (°C)	Conversion (%)	M_n	M_w/M_n
1	**8**	EDC·HCl DMAP	RT	99	16000	1.9
2	**8**	EDC·HCl DMAP	40	99	13000	6.5
3	**8**	EDC·HCl DMAP	RT	59	39600	1.8
4	**9**	TMEDA	RT	49	24000	3.4
5	**9**	TMEDA	40	96	13000	1.9

a) The grafting reaction was carried out using **7** (M_n 19000, M_w/M_n 1.4, θ 72%).

3.3. Investigation of Rotational Behavior of the Components. Longitudinal relaxation times (T_1) measured by ^{13}C NMR for **7** and **10** in $CDCl_3$ at 298 K provided the average time scale for molecular motions of α-CDs. In the observation of the motion of rotaxanes, it is necessary to distinguish the motion of axle polymer and α-CD components. We measured the T_1 for ^{13}C of the 1 and 2-position in α-CDs and ^{13}C of methylene moiety of PTHF axle by using 300 and 400 MHz NMR spectrometers. Table 2 shows the relationship between the obtained T_1 data and the resonance frequency for polyrotaxane. Curve fitting was performed on these plots by using eq. 1 to determine the rotational correlation time (τ_c) showing the molecular fluctuation. The relative movements between α-CDs and axle polymer were estimated by $\Delta(1/\tau_c)$. The τ_c of the axle polymer was smaller than those of α-CDs in each polyrotaxane, indicating that the relative circumrotation rates of axle polymer are faster than that of α-CDs. Because the order of $\Delta(1/\tau_c)$ of α-CDs and/or axle polymer in polyrotaxanes was almost same, it turned out that the wheel components on the graftpolyrotaxane maintained a rotatory capability without a lack of characteristic property of main chain-type polyrotaxane. The results revealed that the movement of the grafting polymer on the wheel components is also independent on the movement of the main chain, strongly suggesting a new type of graft polymer.

Table 2. Relative circumrotation of graftpolyrotaxane.

Rotaxane	τ_c x 10^{11} (s)			$1/\tau_c$ (axle) – $1/\tau_c$ (C1) x 10^{10} (s^{-1})
	axle	C1	C2	
Graftpolyrotaxane **10**	3.40	23.4	43.0	2.51
Polyrotaxane **7**	2.37	41.2	25.4	3.98

4. Summary This paper has disclosed the synthesis, characterization, and dynamic property of graftpolyrotaxane. Grafting reaction based on the graft-onto protocol resulted in the efficient synthesis of graftpolyrotaxane from main chain-type reactive polyrotaxane possessing monofunctional wheels. The present study clearly suggested that the new type of graft polymer, graftpolyrotaxane, would enable the attachment of special characteristics to the graft polymer.

Acknowledgment The present work was financially supported by the Global COE program (Tokyo Institute of Technology) and a Grant-in-Aid for Scientific Research (JSPS).

REFERENCES

[1] For selected reviews concerning polyrotaxane and polyrotaxane network, see: (a) Gibson H W, and Marand H, 1993, *Adv. Mater.* **5**, 11 (b) Sauvage J-P, and Dietrich-Buchecker C. 1999, *Molecular Catenanes, Rotaxanes, and Knots*, Wiley-VCH, New York, Weinheim. (c) Huang F, and Gibson H W, 2005, Polypseudorotaxanes and Polyrotaxanes, *Progr. Polym. Sci.*, **30**, 982 (d) Huang F, Pederson A M, and Gibson H W, 2007, in *Physical Properties of Polymers Handbook*; Mark, J. E., ed.; Springer–Verlag: New York, 693 (e) Kihara N, and Takata T, 2001, *J. Synth. Org. Chem. Jpn.* 59, 206 (f) Takata T, Kihara N, and Furusho Y 2004, *Adv. Polym. Sci.* **171**, 1 (g) Takata T, 2006, *Polym. J.* **38**, 1

[2] (a) Fujita H, Ooya T, and Yui N, 1999, *Macromolecules*, **32**, 2534 (b) Okumura Y, and Ito K, 2001, *Adv. Mater.* **13**, 485 (c) Oku T, Furusho Y, and Takata T, 2004, *Angew. Chem. Int. Ed.* **43**, 966

[3] Yang C, Wang X, Li H, Goh S H, Li J, 2007, *Biomacromolecules*, **8**, 3365

[4] (a) Nayak R R, Swain J R, Mishra M K, and Lenka S, 2004, *Int. J. Plast. Tech.* **8**, 257 (b) Araki J, Kataoka, T, and Ito K, 2008, *Soft Matter.* **4**, 245

[5] For a related report using [2]rotaxane, see: Nishimura D, Oshikiri T, Takashima Y, Hashidzume A, Yamaguchi H, and Harada A, 2008, *J. Org. Chem.* **73**, 2496.

[6] (a) Melton L D, and Slessor K N, 1971, *Carbohydr. Res.* **18**, 29 (b) Tanaka M, Kawaguchi Y, Niinae T, and Shono T, 1984, *J. Chromotogr.* **314**, 193 (c) Kaneda T, Fujimoto T, Goto J, Asano K, Yasufuku Y, Jung J H, Hosono C, and Sakata Y, 2002, *Chem. Lett.* **31**, 514

[7] (a) Choi J, Kwon S, and Mah S, 2002, *J. Appl. Polym. Soc*, **83**, 2082 (b) Shimomura O, Lee B, Meth S, Suzuki H, Mahajan S, Nomura R, and Janda K, 2005, *Tetrahedron*, **61**, 12160 (c) Ouis N, Benharrats N, and Belbachir M, 2004, *C. R. Chimie*, **7**, 955 (d) Ing H, and Manskr R, 1926, *J. Chem. Soc*, 2348

[8] Hardy D V N, 1934, *J. Chem. Soc.* 2001

[9] Kihara N, Hinoue K, and Takata T, 2005, *Macromolecules*, **38**, 223

[10] Arai T, and Takata T, 2007, *Chem. Lett.* **36**, 413

Journal of Physics: Conference Series **184** (2009) 012025 doi:10.1088/1742-6596/184/1/012025

Metal-Catalyzed Living Radical Polymerization and Radical Polyaddition for Precision Polymer Synthesis

M Mizutani, K Satoh and M Kamigaito [1]

Department of Applied Chemistry, Graduate School of Engineering, Nagoya University, Furo-cho, Chikusa-ku, Nagoya 464-8603, Japan

E-mail: kamigait@apchem.nagoya-u.ac.jp

Abstract. The metal-catalyzed radical addition reaction can be evolved into two different polymerization mechanisms, i.e.; chain- and step-growth polymerizations, while both the polymerizations are based on the same metal-catalyzed radical formation reaction. The former is a widely employed metal-catalyzed living radical polymerization or atom transfer radical polymerization of common vinyl monomers, and the latter is a novel metal-catalyzed radical polyaddition of designed monomer with an unconjugated C=C double bond and a reactive C–Cl bond in one molecule. The simultaneous ruthenium-catalyzed living radical polymerization of methyl acrylate and radical polyaddition of 3-butenyl 2-chloropropionate was achieved with $Ru(Cp^*)Cl(PPh_3)_2$ to afford the controlled polymers, in which the homopolymer segments with the controlled chain length were connected by the ester linkage.

1. Introduction

Recent remarkable progress in controlled/living radical polymerizations has enabled the precision polymer synthesis of a wide variety of well-defined polymers, such as end-functionalized, block, graft, and star polymers [1–4]. Among them, the metal-catalyzed living radical polymerization or atom transfer radical polymerization (ATRP) is one of the most efficient systems for various conjugated vinyl monomers, such as methacrylates, acrylates, acrylamides, and styrenes [2, 3]. This polymerization was originally developed from the metal-catalyzed Kharasch or atom-transfer radical addition reaction (ATRA), in which the carbon radical species is generated through the metal-catalyzed cleavage of a carbon–halogen (C–X) bond and adds to an olefin to form the 1:1 adduct of the halide and olefin [5, 6]. The metal-catalyzed living radical chain-growth polymerization proceeds via the metal-catalyzed reversible formation of the growing radical species from the dormant C–X polymer terminal, and the addition of the growing radical species to the vinyl monomers [7].

In contrast, we quite recently found a novel radical polyaddition reaction of a designed monomer that possesses an active C–X bond and unconjugated C=C double bond via the evolution of the metal-catalyzed ATRA into the step-growth polymerization [8, 9]. In this polymerization, the active or dormant C–X bond in the monomer is activated by the metal catalysts to form a radical species, which adds to the C=C double bond of another monomer molecule to generate a C–C bond as the main chain, along with an inactive C–X bond as the pendant. This radical polyaddition can be catalyzed by the same metal catalyst for the living radical polymerization.

[1] To whom any correspondence should be addressed.

This short paper will first describe the metal-catalyzed chain-growth living radical polymerization of methyl acrylate (MA) and then the metal-catalyzed step-growth radical polyaddition of a designed monomer (**1**), in which the unconjugated C=C and reactive C–C bonds are linked via ester linkage. Finally, the simultaneous metal-catalyzed radical living polymerization of MA and radical polyaddition of **1** was examined for novel precision polymer synthesis (scheme 1).

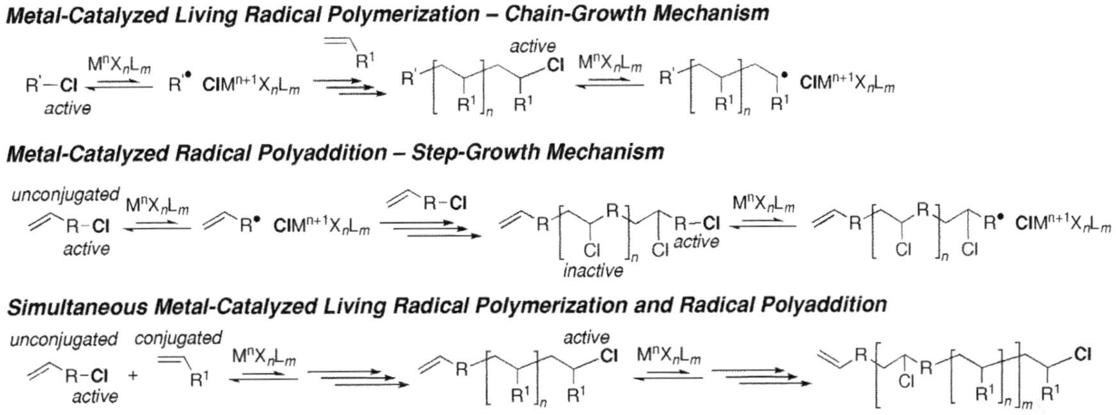

Scheme 1. Metal-catalyzed living radical polymerization and radical polyaddition

2. Experimental

$Ru(Cp^*)Cl(PPh_3)_2$ (provided from Wako) and CuCl (Aldrich, 99.99%) were used as received. All metal compounds were handled in a glovebox (VAC Nexus) under a moisture- and oxygen-free atmosphere (O_2, < 1 ppm). Toluene was distilled over sodium benzophenone ketyl and bubbled with dry nitrogen for 15 min just before use. Methyl acrylate (TCI, >99%), methyl 2-chloropropionate (TCI, >95%), *N,N,N',N'',N''*-pentamethyldiethylenetriamine (PMDETA) (TCI, >98%), and *n*-Bu_3N (Wako, >98%) were distilled from calcium hydride under reduced pressure before use. 3-Butenyl 2-chloropropionate (**1**) was synthesized from 2-chloropropionyl chloride (TCI, >95%) and 3-buten-1-ol (TCI, >98%).

Polymerization was carried out under dry nitrogen in baked glass tubes equipped with a three-way stopcock using syringe techniques.

Monomer conversion was determined from the concentration of residual monomer measured by gas chromatography. Functional group (C=C and C–Cl) conversions were measured by 400 MHz ^{1}H NMR (Varian Gemini 2000). The number-average molecular weight (M_n), the weight-average molecular weight (M_w), and the molecular weight distribution (MWD) of the polymers were determined by size-exclusion chromatography (SEC) in THF. The columns were calibrated against eight standard poly(methyl methacrylate) samples (Shodex; M_p = 202–1,950,000; M_w/M_n = 1.02–1.09).

3. Results and discussion

3.1. Metal-catalyzed radical living polymerization of MA

First, we investigated the ruthenium-catalyzed living radical polymerization of MA using $Ru(Cp^*)Cl(PPh_3)_2$ and methyl 2-chloropropionate (**2**), which has a similar secondary C–Cl bond to **1**, in the presence of *n*-Bu_3N in toluene at 80 °C. The polymerization occurred smoothly (figure 1A) to give the polymers with relatively narrow MWDs (figure 1C). However, the MWDs were slightly broader than those obtained with the methyl methacrylate dimer-type initiator [H–$(MMA)_2$–Cl], which has a more reactive C–Cl bond [10], indicating a slower initiation from **2**. The M_n of the polymers increased in direct proportion to monomer conversion and agreed well with the calculated values, assuming that one molecule of **2** generates one living polymer chain (figure 1B).

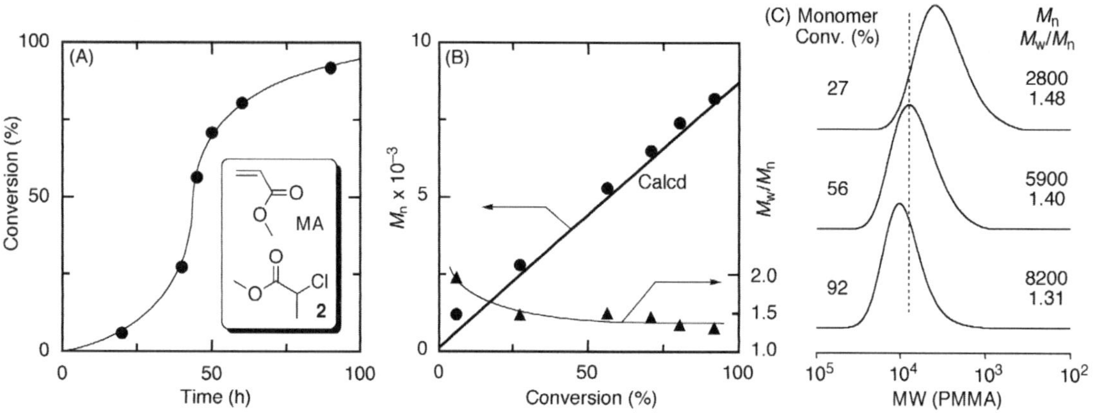

Figure 1. Metal-catalyzed living radical polyaddition of MA with **2**/Ru(Cp*)Cl(PPh_3)$_2$/n-Bu_3N in toluene at 80 °C: [MA]$_0$/[**2**]$_0$/[Ru(Cp*)Cl(PPh_3)$_2$]$_0$/[n-Bu_3N]$_0$ = 4000/40/4/40 mM.

These results indicate that living radical polymerization of MA proceeds via reversible activation of C–Cl bond originating from the chloride initiator **2**.

3.2. Metal-catalyzed radical polyaddition of 1

The metal-catalyzed radical polyaddition of **1** was conducted with CuCl/PMDETA in toluene at 60 °C [11]. The monomers were consumed smoothly and almost completely irrespective of the slow reaction (figure 2A). The consumption of the two functional groups in **1**, C=C and C–Cl, occurred at the same rate and was slower than that of **1**, indicating the 1:1 intermolecular addition reaction. The SEC curves of the obtained polymers shifted to high molecular weights as the polymerization proceeded (figure 2C). The M_n of the products progressively increased and was close to the calculated line for the step-growth polymerization, assuming that M_n increases in inverse proportion to 1-p, where p and MW(**1**) mean the consumption of the functional group and the molecular weight of the monomer (**1**) (figure 2B) [9]. The MWD increased with conversion and was also close to the calculated value of 2 for step-growth polymerization.

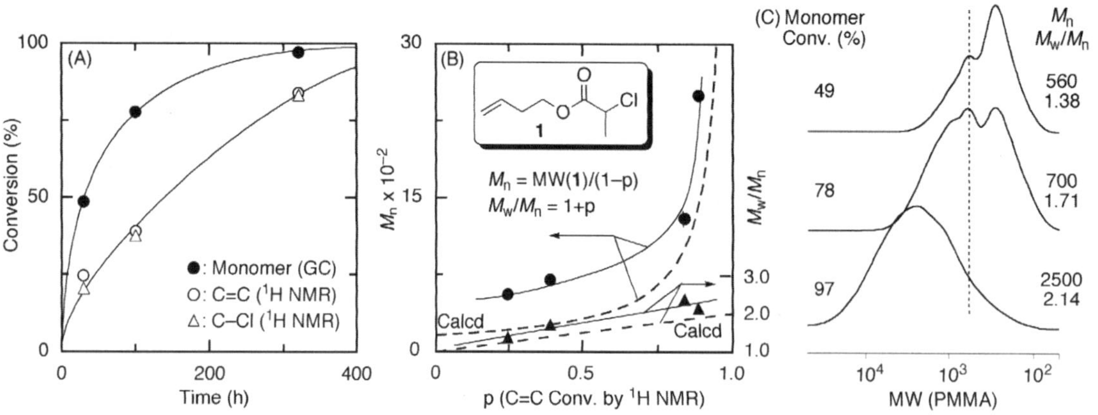

Figure 2. Metal-catalyzed radical polyaddition of **1** with CuCl/PMDETA in toluene at 60 °C: [**1**]$_0$/[CuCl]$_0$/[PMDETA]$_0$ = 2000/100/400 mM.

Thus, the polymerization of **1** proceeds via the expected intermolecular radical polyaddition, which is based on the metal-catalyzed C–Cl activation similar to the metal-catalyzed living radical polymerization.

3.3. Simultaneous metal-catalyzed radial living polymerization of MA and polyaddition of ***1***

The simultaneous metal-catalyzed living radical polymerization of MA and radical polyaddition of **1** ($[MA]_0/[\mathbf{1}]_0$ = 4000/40 mM) was then investigated by the ruthenium catalyst in toluene at 80 °C. As shown in figure 3A, both the monomers were consumed simultaneously, where the consumption of **1** was faster than that of MA. During the early stage of the polymerization, the M_n of the obtained polymers increased in direct proportion to the conversion of MA and agreed well with the calculated value, assuming that one molecule of **1** generates one polymer chain (figure 3B). Furthermore, the SEC curves were relatively narrow and shifted to high molecular weights along with the consumption of MA (figure 3C). These results indicate that living radical polymerization of MA first proceeds via reversible activation of the C–Cl terminal derived from the C–Cl bond of **1** via the metal-catalyzed initiation. However, in the later stage, the molecular weight of the polymers drastically increased along with change of the unimodal SEC curves into multimodal, indicating that the polyaddition reaction occurred between the living polymer chains possessing C=C and C–Cl groups at the α- and ω-chain ends, respectively, both of which originated from **1**. Due to the lower reactivity of the unconjugated double bond of **1**, the polyaddition reaction was slower than the addition polymerization of MA and resulted in a progressive increase of the molecular weight in the later stage of the reaction.

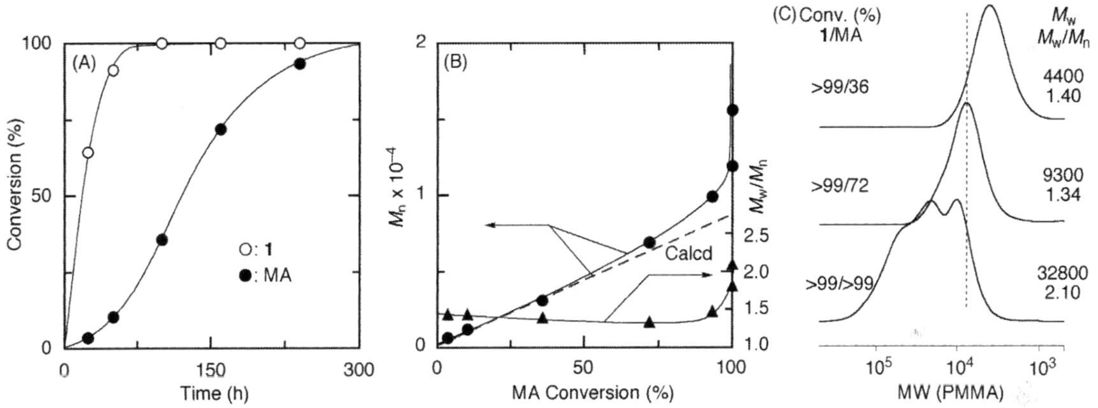

Figure 3. Simultaneous metal-catalyzed living radical polymerization of MA and radical polyaddition of **1** with $Ru(Cp^*)Cl(PPh_3)_2$/n-Bu_3N in toluene at 80 °C: $[MA]_0/[\mathbf{1}]_0/[Ru(Cp^*)Cl(PPh_3)_2]_0/[n\text{-}Bu_3N]_0$ = 4000/40/4/40 mM.

In conclusion, the simultaneous radical living polymerization and radical polyaddition occur via the same metal-catalyzed radical formation reaction to afford the controlled polymers, in which the homopolymer segments with the controlled chain length were connected by the ester linkage.

References

[1] Hawker C J, Bosman A W and Harth E 2001 *Chem. Rev.* **101** 3661
[2] Kamigaito M, Ando T and Sawamoto M 2001 *Chem. Rev.* **101** 3689
[3] Matyjaszewski K and Xia J 2001 *Chem. Rev.* **101** 2921
[4] Moad G, Rizzardo E and Thang S H 2005 *Aust. J. Chem.* **58** 379
[5] Kharasch M S, Jensen E V and Urry W H 1945 *Science* **102** 128
[6] Iqbal J, Bhatia B and Nayyar N K 1994 *Chem. Rev.* **94** 519
[7] Kato M, Kamigaito M, Sawamoto M and Higashimura T 1995 *Macromolecules* **28** 1721
[8] Satoh K, Mizutani M and Kamigaito M 2007 *Chem. Commun.* 1260
[9] Mizutani M, Satoh, K and Kamigaito M 2009 *Macromolecules* **42** 472
[10] Watanabe Y, Ando T, Kamigaito M and Sawamoto M 2001 *Macromolecules* **34** 4370
[11] Xia J and Matyjaszewski K 1997 *Macromolecules* **30** 7697

Journal of Physics: Conference Series **184** (2009) 012026 doi:10.1088/1742-6596/184/1/012026

Residual Stress of Particulate Polymer Composites with Reduced Thermal Expansion

T Nishino, M Kotera, Y Sugiura

Department of Chemical Science and Engineering, Graduate School of Engineering, Kobe University, Rokko, Nada, Kobe 657-8501, Japan
E-mail: tnishino@kobe-u.ac.jp

Abstract. Thermal expansion behavior was investigated for tangusten zirconium phosphate ($Zr_2(WO_4)(PO_4)_2$ (ZWP)) particulate filled poly(ether ether ketone) (PEEK) composite. ZWP is known as ceramic filler with a negative thermal expansion. By incorporating ZWP with 40 volume %, the linear thermal expansion coefficient of the PEEK composite was reduced to almost same value ($2.53\ X10^{-5}\ K^{-1}$) with that of aluminum. This decrease was found to be quite effective for the decrease of the residual stress at the interface between aluminum plate and the composite.

1. Introduction

Thermal expansion behavior is one of the most important thermal properties of the composites. The linear thermal expansion coefficients α of conventional polymers are roughly in the range of $10^{-5}\ K^{-1}$, which is far higher than those of metals and ceramics [1]. Assuming the case that polymer was adhered to metal or ceramics, the mismatch of the thermal expansion will induce residual stress at the interface. This residual stress often damages the products seriously through debonding for the adhesives, bending and delamination for the laminates, leak by water uptake for IC, LSI devices. Thus the decrease of the residual stress is very important from the practical usages [2]. The thermally induced residual stress is well known as functions of i) elastic modulus, ii) glass transition temperature and iii) thermal expansion of the material [3]. Thus, the decreases of any of these three factors will contribute to the decrease of the residual stress. However, the former two factors are very important from the mechanical and thermal performance of the materials. Accordingly, the last factor, *that is*, the decrease of the thermal expansion is the most preferable one to fulfill the purpose.

The thermal expansion behavior of the composite can be controllable by changing the combination of the incorporated fillers/fibers and the matrix, volume fraction, orientation, dispersion and so on. Some of the ceramics are known to show negative thermal expansion. For example, ZrW_2O_8 [4],$PbTiO_3$ [5] are found to show negative α in a wide temperature range. Thus the researchers synthesized micro- and nano-size particles, then incorporated these particles into polymer. They succeeded to reduce the α values of the polymeric composites for epoxy resin [6], poly(methyl methacrylate) [7], polyimide [8], cyanate ester [9]. Zirconium tungsten phosphate, $Zr_2(WO_4)(PO_4)_2$ (ZWP) also shows negative α of - $3.2\ X\ 10^{-6}\ K^{-1}$, which is available commercially.

In this study, we report the particulate polymer composite with reduced thermal expansion, by combining ZWP particles with poly(ether ether ketone) (PEEK). PEEK has found much application as super engineering plastic, where thermal dimensional stability with low α value is highly required. In addition to the thermal expansion behavior, the residual stress of the composite was evaluated.

Journal of Physics: Conference Series **184** (2009) 012026 doi:10.1088/1742-6596/184/1/012026

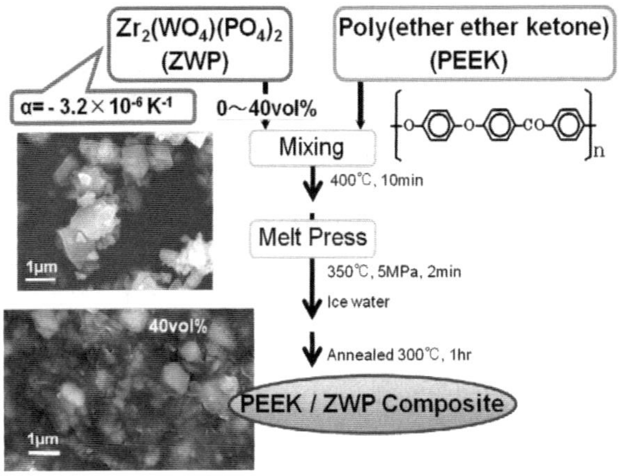

Figure 1 Sample preparation scheme of PEEK /ZWP composites.

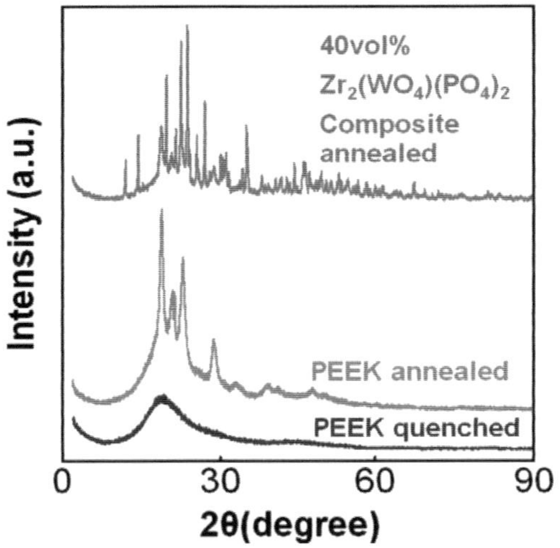

Figure 2 X-ray diffraction profiles of PEEK and PEEK/ZWP composite (40 vol%) .

2. Experimental

2.1 Sample Preparation

Commercial available ZWP particles, kindly supplied from KCM Co., Japan, was mixed with PEEK (Sumitomo Chemical Co.Ltd., 450 G) using a brabender type mixer (Labo Plastomill, Toyo Seiki Seisaku-sho Ltd.) at 400 °C for 10 min. Then the mixture was melt pressed into the sheet shaped composite at 350 °C, 5 MPa for 2 min, followed by quenching into ice water (Quenched). After that, the composite was annealed at 300 °C for 1 h (Annealed). The thickness of the composite was 0.3 mm. The sample preparation scheme is shown in Figure 1 with scanning electron micrographs.

2.2 Measurements

The structure of the composites was investigated by X-ray diffraction using Cu Kα radiation, generated with RINT-2000 (RIGAKU) at 40 kV, 20 mA. Thermomechanical analysis, TMA, was performed using a themomechanical analyzer (SSC/5200 TMA 120 CU, Seiko Instruments). The tensile stress of 0.1 MPa was applied to the rectangular sample with its original length of 15 mm. Then the sample was heated at 10 °C/min. The thermal expansion coefficient was evaluated from the inclination of the strain vs. temperature curve. The filler volume fraction in the composite was evaluated from the thermal gravimetric trance with the heating rate of 10 °C/min under nitrogen flow. The ZWP particles and the fracture surface of the composite was observed using a scanning electron microscope (SEM) (JSM-5610LVS (JEOL)), at an accelerating voltage of 20 kV. Pt/Pd was deposited on the surface prior to the observation.

3. Results and Discussion

3.1 Structure of PEEK / ZWP Composites

The ZWP particles were crushed shape and the size was widely distributed from 0.1 to several μ m. The ZWP particles were homogeneously dispersed in the PEEK matrix as shown in Fig.1.

Figure 2 shows the X-ray diffraction profiles of quenched, annealed PEEK and annealed PEEK / ZWP (40 vol %) composite. The quenched PEEK is amorphous, however, it crystallized after annealing. In addition, the diffraction peaks assigned as ZWP were overlapped to the amorphous halo and the crystalline peaks from PEEK for the composite. These reveal that the crystalline state of the composites changed drastically by annealing.

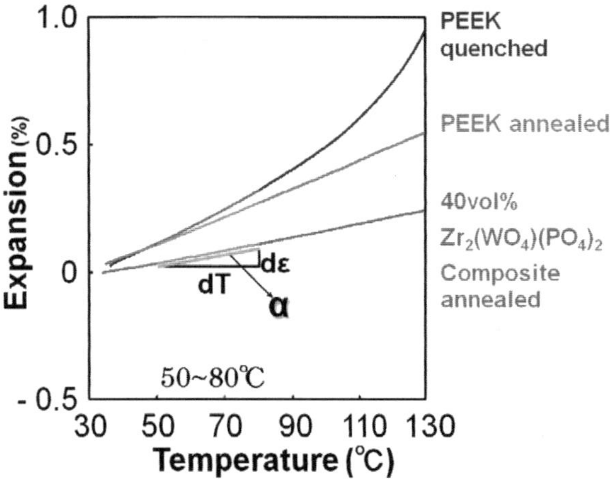

Figure 3 Relationship between the thermal strain and temperature of. quenched, annealed PEEK and annealed PEEK / ZWP (40 vol %) composite.

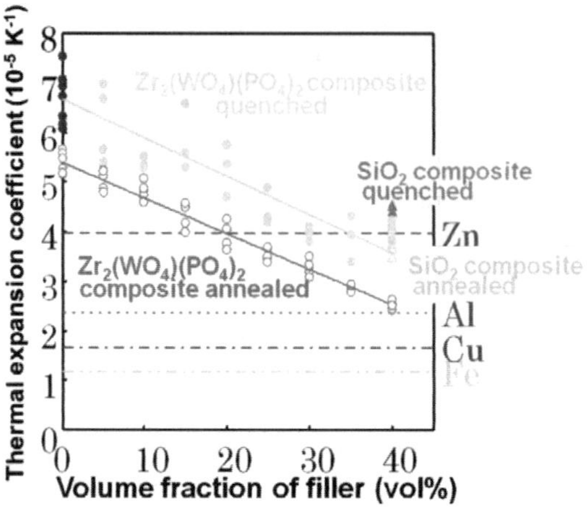

Figure 4 Relationship between the thermal expansion coefficient and he volume fraction of ZWP for the quenched and annealed ZWP particulate PEEK composites.

3.2 Thermal Properties and Residual Stress of PEEK/ZWP Composites

Figure 3 shows the relationship between the thermal strain and temperature of the PEEK films and the composite from the TMA measurements. With increasing temperature, all of them showed thermal expansion. Especially, the quenched PEEK film showed nonlinear large expansion due to the amorphous state. After annealing, the inclination of the curve decreased, then by incorporating ZWP, it decreased further. Here, the α value is defined from the linear part of the curve between 50 and 80°C as shown in the figure.

Figure 4 shows the relationship between the α value and the volume fraction of ZWP for the quenched and annealed ZWP particulate PEEK composites. The α values of some metals and PEEK/SiO_2 (40 vol %) composite were also superimposed on the figure. With increasing the volume fraction of ZWP particles, the α value decreased linearly. The absolute α value of the annealed composite was lower than those of the quenched ones. This is reflected by the α value of each matrix. Above 20 vol % for the annealed composite (corresponding to the 35 vol % for the quenched composite), the α values were lower than that of Zn, and it reached to that of Al (2.36 X10^{-5} K^{-1}) at 40 vol % for the annealed composite. This value is far lower than that of SiO_2 particulate filled composites, in which the α value of SiO_2 is 1.3 X $10^{-5}K^{-1}$. Thus the incorporation of the particles with low (negative) α value is quite effective for lowering the thermal expansion of the composite. As the first–order approximation to the overall calculation, the calculated α_{calcd} value can be evaluated using the simple rule of mixture as follows:

$$\alpha_{calcd} = \alpha_f \phi + \alpha_m (1-\phi)$$

where, ϕ is the volume fraction of filler, and α_f, α_m are the α values of the filler and the matrix, respectively. The α_{calcd} vales for the quenched and annealed composites with ϕ=40 vol % are 4.2 X$10^{-5}K^{-1}$, 3.3 X$10^{-5}K^{-1}$, respectively, which were higher than the observed values. The equation simply predicts the α value without considering dispersivity of the ZWP particles in the matrix and/or the interfacial interaction between ZWP and PEEK. Further investigations will be needed to reproduce the observed ones. The extrapolating of the red line for the annealed composite to the volume fraction of 60%, the α value reached to that of Fe. However, the volume fraction of 40 % of ZWP was the maximum attained value by the brabender mixing so far we investigated. Further improvement of the dispersivity of the particles should be considered.

Figure 5 shows the photographs of the laminates composed of (upper) quenched PEEK sheet / Al plate, and (lower) annealed PEEK / ZWP (40 vol %) composite / Al plate. The PEEK sheet and the composite sheet were adhered to the Al plate (thickness: 0.3 mm) using the epoxy resin, respectively, then cured at 140 °C followed by rapidly cooling to room temperature.

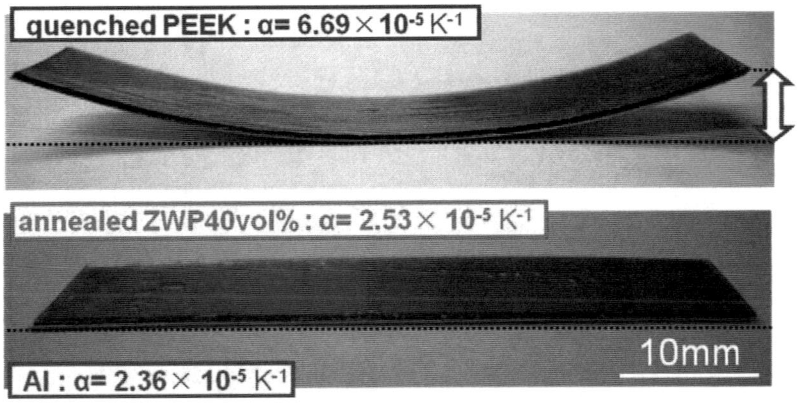

Figure 5 Photographs of the laminates composed of (upper) quenched PEEK sheet / Al plate, and (lower) annealed PEEK / ZWP (40 vol %) composite / Al plate.

The laminate of quenched PEEK and Al was bended with the PEEK sheet inside. This is because the thermal shrinkage of the PEEK film is higher than that of the Al plate during cooling from the curing temperature to room temperature. It is clear from these photographs, the radius of the curvature decreased for the composite. This reveals that the residual stress could be reduced by the decreased of the thermal expansion of the polymer composite. The residual stress $\sigma_{calcd.}$ can be calculated as 20.1 MPa , 1.4 MPa for the quenched PEEK and the annealed composite, respectively, by considering the elastic modulus of the quenched PEEK sheet (1.7 GPa) and the annealed composite (4.1 GPa) [10]. This corresponds to the observed ones shown in Fig.5.

4. Conclusions

By incorporating ZWP particles with negative thermal expansion, the thermal expansion coefficient of the PEEK composite could be reduced to almost same value with that of Al. These decreases of the thermal expansion coefficient were quite effective for the decrease of the residual stress at the interface.

References

[1] Hull D and Clyne T W 1996 *An Introduction to Composite Materials* (Cambridge Univ. Press.)
[2] Ramani K and Zhao W 1997 *Intern'l J. Adhesion and Adhesives* **17** 353
[3] Ochi M, Yamashita K and Shimbo M 1991 *J.Appl.Polym.Sci.* **43** 2013
[4] Mary T A, Evans J S O, Vogt T and Sleight A W 1996 *Science* **272** 90
[5] Chamdra A, Pandey D, Mathews M D and Tyagi A K 2005 *J.Mater.Res.* **20** 350
[6] Wong C P and Bollampally R S 1999 *J.Appl.Polym.Sci.* 74 3396
[7] Chamdra A, Meyer W H, Best A, Hanewald A, Wegner G 2007 *Macromol.Mater.Eng.* **292** 295
[8] Sullivan L M and Lukehart C M 2005 *Chem.Mater.* **17** 2136
[9] Goertzen W K and Kessler M R 2008 *J.Therm.Anal.Cal.* **93** 87
[10].Nakamae K, Nishino T, Xu A, Matsumoto T and Miyamoto T 1990 *J.Appl.Polym.Sci.* **40** 2231

Field Emission Scanning Electron Microscopy Combined with Focused Ion Beam for Rubbery Material with Nano-Matrix Structure

Keiichi Akabori[1,2], Yoshimasa Yamamoto[1,2], Seiichi Kawahara[1,2,5]
Hiroshi Jinnai[3], and Hideo Nishioka[4]

[1]Department of Materials Science and Technology, Faculty of Engineering, Nagaoka University of Technology, Nagaoka 940-2188, Japan,
[2]Center for Green-Tech Development in Asia,
[3]Department of Macromolecular Science and Engineering, Graduate School of Science and Engineering, Kyoto Institute of Technology, Matsugasaki, Kyoto 606-8585, Japan
[4]JEOL Ltd., 1-2-3 Musashino, Akishima, Tokyo 196-8556, Japan

kawahara@mst.nagaokaut.ac.jp

Abstract. Nano-matrix structure of a graft-copolymer consisting of deproteinized natural rubber (DPNR) and polystyrene (PS) was observed through field-emission scanning microscopy combined with a focused gallium (Ga) ion beam (FIB-SEM). The graft-copolymer was carefully stained with osmium tetraoxide for more than a week to avoid a charge-up through FIB-SEM. Field emission SEM images of the graft-copolymer were taken after etching its surface with the Ga ion at each interval and they were reconstructed to obtain the three-dimensional image of the nano-matrix structure. The NR particles of about 0.5 μm in diameter were found to be dispersed into the nano-matrix of PS, which connected to each other on all directions. The dimensional feature of the NR particles was dependent upon an expression of orthogonal views of the 3D image.

1. Introduction

Field emission scanning electron microscopy (FE-SEM) combined with a focused Ga ion beam (FIB) is a powerful technique to observe a three-dimensional structure of nano-phase-separated materials at very low magnification.[1] We are, thus, able to know an entire image of the phase-separated structure, which is closely related to mechanical properties of the materials. The three-dimensional image is made by reconstruction of a number of FE-SEM images, which are taken at each a hundred or less nm interval after etching a surface of the materials with FIB. In this case, a resolution of the three-dimensional image in Z-direction is governed by a length between the etched surfaces, whereas resolutions in X and Y-directions are dependent upon an ability of FE-SEM to obtain a fine image. To take a high resolution three-dimensional image, less length between the etched surfaces must be accomplished with FIB.

[5] Author to whom all any correspondence should be addressed.

In the previous work,[1] Jinnai and co-workers observed a three-dimensional lamellar structure of a polystyrene-*block*-polyisoprene through FIB-SEM. The three-dimensional lamellar structure with an exceptionally large periodicity of about 0.1 μm was well-reconstructed in a FIB-SEM image. On the other hand, we observed a three-dimensional nano-matrix structure of deproteinized natural rubber-*graft*-polystyrene (DPNR-*graft*-PS) through FIB-SEM.[2] The nano-matrix structure was found to be a unique nano-phase-separated structure, which consisted of DPNR dispersoid of about 0.5 μm in diameter and PS matrix of about 15 nm in thickness. In order to analyze the nano-matrix structure, more exactly, it is necessary to observe the structure in various directions, i.e X-Y plane, X-Z plane, Y-Z plane and so forth.

In the present work, we performed a three-dimensional observation of the nano-matrix structure, which is formed by graft-copolymerization of styrene onto deproteinized natural rubber in latex stage followed by casting a film from the reacted latex.[3] Almost all polystyrene of about 10^4 g/mol in number average molecular weight, M_n, is chemically attached to deproteinized natural rubber (M_n: ca. $3x10^5$ g/mol) particle at the interface, as is evident from high grafting efficiency of styrene, i.e. more than 90 mol%. The resulting film is carefully stained with osmium tetraoxide (OsO_4) and it is subjected to FIB-SEM.

2. Experimental

Natural rubber (NR) latex used in this study was commercial high-ammonia NR latex. The incubation of the latex was made with 0.1 w/w% urea in the presence of 1 w/w% sodium dodecyl sulfate (SDS) at 303K followed by centrifugation at 10,000 g.[4] The cream fraction was re-dispersed in 1 w/w% SDS solution to make latex of 30 w/w% in dry rubber content (DRC) and it was washed twice by centrifugation to prepare deproteinized natural rubber (DPNR) latex. The DPNR latex was diluted with distilled water to adjust 30 w/w% DRC and concentration of SDS was adjusted to 0.1 w/w%.

Graft-copolymerization of styrene onto DPNR was carried out in latex stage with *tert*-butyl hydroperoxide / tetraethylene pentamine as an initiator.[3] The DPNR latex was charged with N_2 gas for 1 hour at 30℃. The initiator of 3.3×10^{-2} mol/kg-rubber and styrene of 1.5 mol/kg-rubber were added in turn to the latex. The reaction was performed by stirring the latex at 400 rpm for 2 hours at 30℃. Unreacted styrene was removed with a rotary evaporator under reduced pressure. A film specimen of the resulting NR (DPNR-*graft*-PS) was prepared by casting from the reacted latex in a Petri dish and it was dried under reduced pressure at an ambient temperature for more than a week.

A three-dimensional (3D) FIB-SEM image was obtained with a SII SMI-3050SE at accelerating voltage of 3 kV, in which positively charged Ga ion at about 50 pA was used for etching process. To prevent the surface from curving during the process, a thin carbon layer was deposited at the surface of the specimen. Field emission SEM images of the DPNR-*graft*-PS were taken after etching its surface at each 140 nm intervals with Ga^+ ion. The 3D image with a pixel size of X=15 nm, Y=19 nm and Z=140 nm and/or 102 nm was made by stacking two-dimensional FE-SEM images. To increase the pixel size at Z-direction as it looks good, the 6 same images were stacked in each interval, so that the pixel size of completed 3D image was improved by a factor of 6 times in Z axis.

3. Results and Discussion

Figure 1 (a) shows FE-SEM image of the etched surface of the DPNR-*graft*-PS. The image was taken in an oblique direction by FE-SEM, since the film specimen, placed on a flat plate in a chamber, was perpendicularly etched with FIB. In the image, an edge of the etched surface was situated rather front compared with a bottom bed; hence, the magnification of the front edge is a little bit higher than that of the bottom bed. This may be explained to be due to the oblique angle between an electron-gun and sample surface. Figure 1 (b) shows a part of the two-dimensional FE-SEM image of the DPNR-*graft*-PS, in which a gloomy domain is assigned to natural rubber (NR) and a bright domain is to polystyrene (PS). The bright PS nano-matrix was quite distinct near the top of the image, whereas this turned into ambiguous near the bottom. This is attributed to a difference between distances of secondary electron detector from top and bottom of the etched surface, due to the oblique angle; the lower the height from the top of the etched surface along to the Y-axis, the worse was the focus. Thus, the scanning region was limited in Y-direction, resulting in the rectangular image in Figure 1 (b). The image was almost similar to that obtained by transmission electron microscopy (TEM),[2] suggesting that the structure of the DPNR-*graft*-PS was not damaged by FIB etching.

It is well-known that the high intense FIB destroys the phase-separated structure of polymeric materials due to the local heating by charging up a specimen during FIB etching.[5] The damage may be more significant for rubbery materials than thermoplastics, since the former possesses lower glass transition temperature and less electric conductivity. To prevent the damage of the DPNR-*graft*-PS during FIB etching, in this work, the DPNR-*graft*-PS was stained with OsO_4 for more than 7 days. This enabled to attain a suitable etching of the DPNR-*graft*-PS.

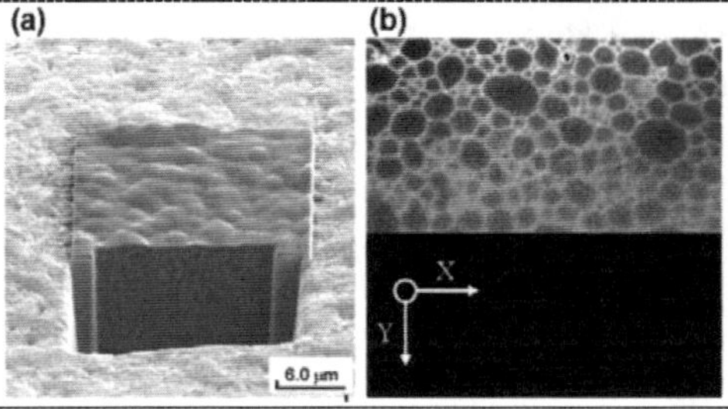

Figure 1. (a) FE-SEM image of the DPNR-*graft*-PS in the etching process and (b) the two-dimensional X-Y image of the etched surface.

Figure 2 shows 3D images of the DPNR-*graft*-PS; (a) etching interval (d) = 140 nm and (b) d = 102 nm. The box size is about X = 6 μm, Y = 4 μm, Z = 7 μm for the image (a) and X = 6 μm, Y = 4 μm, Z = 5 μm for the image (b). The nano-matrix structure of the DPNR-*graft*-PS was well-visualized in the 3D image, since the box size was much larger than the size of the NR particle. The bright PS matrix was found to connect to each other in all directions. The NR particles of about 0.5 μm in diameter were found to be dispersed into the nano-matrix of PS. The average diameter of the NR particles was more than 5 times as large as the etching interval. Thus, a small difference in the etching interval was negligible for a spatial resolution of the 3D images.

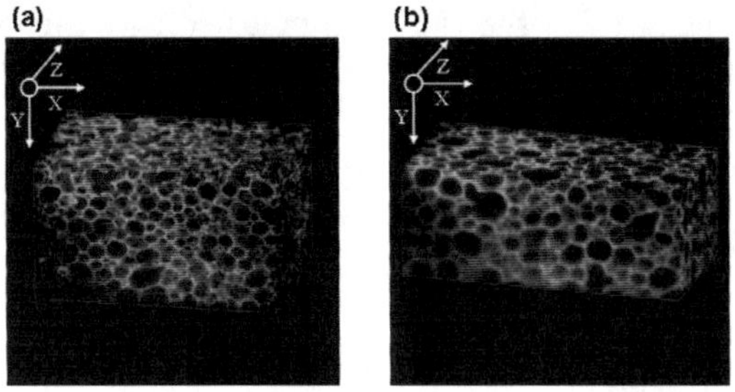

Figure 2. 3D construction images of the DPNR-*graft*-PS; (a) d=140 nm and (b) d=102 nm.

Figure 3 shows typical orthogonal views of 3D image of the DPNR-*graft*-PS after reconstructing the FE-SEM images, where the figures (b), (c) and (d) are corresponding to X-Y, X-Z and Y-Z planes, respectively. The resolution of X-Y plane was quite good due to the high resolution of the FE-SEM. However, the resolution of X-Z and Y-Z planes was reduced. The low resolution of X-Z and Y-Z planes may be explained to be due to the larger etching interval, i.e. approximately 100 nm. This was much larger than 20 nm in the etching interval for the PS-*block*-PI.[1] It was quite difficult to reduce the etching interval for the DPNR-*graft*-PS, since the nano-matrix structure of the DPNR-*graft*-PS was easily damaged by FIB etching compared with the lamellae structure of the PS-*block*-PI, due to larger volume fraction of the rubber components of the DPNR-*graft*-PS.

Size of the NR particles, shown in Figure 3, was analyzed by an ordinary image processor to estimate an average diameter. The estimated values of the average diameter in the long axis were approximately 0.55±0.38 μm in X-Y plane, 0.41±0.12 μm in X-Z plane and 0.58±0.25 μm in Y-Z plane, respectively. The average diameter in the X-Z plane was smaller than those in the X-Y and Y-Z planes. Furthermore, the NR particles in the X-Y and Y-Z planes resembled to be ellipsoidal, which is different from circular NR particles observed by transmission electron micro-tomography.[2] If the morphology of the DPNR-*graft*-PS have been isotropic, the average diameter of the NR particles would be equal to each other in orthogonal views of the 3D image. These unexpected results suggest that the difference in the average diameter in the each plane is dependent upon a detection system of FE-SEM, where an electron-gun is obliquely arranged to the sample stage. To perform the quantitative analysis, more precisely, further improvement must be required for not only the computational graphics but also a novel etching technique. These are now being conducted and will be reported in our subsequent paper.

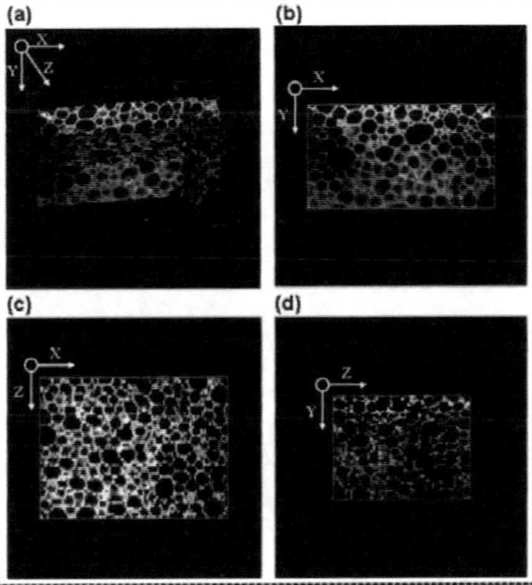

Figure 3. Typical orthogonal views of FIB-SEM image of the DPNR-*graft*-PS after the 3D construction.

4. Conclusions

The 3D visualization of the nano-matrix structure of the DPNR-*graft*-PS was performed in the micrometric scale by FIB-SEM. The DPNR-*graft*-PS strained with OsO_4 for more than 7 days was etched by FIB at the interval of approximately 100 nm without any damage during FIB etching. The resolution of the X-Y plane in the 3D image was quite good due to the high resolution of the FE-SEM, whereas the resolution of X-Z and Y-Z planes was reduced due to the larger etching interval of approximately 100 nm. The estimated values of the average diameter of the NR particles were dependent upon the oblique geometry of FE-SEM, especially with respect to Y-direction.

References

[1] Kato M, Ito T, Aoyama Y, Sawa K, Kaneko T, Kawase N and and Jinnai H 2007 *J. Polym Sci: Part B: Polym. Phys.* **45** 677
[2] Kawahara S, Yamamoto Y, Fujii S, Isono Y, Niihara K, Jinnai H, Nishioka H and Takaoka A 2008 *Macromolecules* **41** 4510
[3] Kawahara S, Kawazura T, Sawada T and Isono Y 2003 *Polymer* **44** 4527
[4] Kawahara S, Klinklai W, Kuroda H and Isono Y 2004 *Polym. Adv. Technol.* **15** 181
[5] Niihara K, Kaneko T, Suzuki T, Sato Y, Nishioka H, Nishi T and Jinnai H 2005 *Macromolecules* **38** 3048

Journal of Physics: Conference Series **184** (2009) 012028 doi:10.1088/1742-6596/184/1/012028

Chain and Mirophase-separated Structures of Ultrathin Polyurethane Films

Ken Kojio[1], Yusuke Uchiba[2], Yasunori Yamamoto[1], Suguru Motokucho[1], Mutsuhisa Furukawa[1,2,3]

[1]Faculty of Engineering, [2]Graduate School of Materials and Science, Nagasaki University, 1-14 Bunkyo-machi, Nagasaki 852-8521, Japan

furukawa@nagasaki-u.ac.jp

Abstract. Measurements are presented how chain and microphase-separated structures of ultrathin polyurethane (PU) films are controlled by the thickness. The film thickness is varied by a solution concentration for spin coating. The systems are PUs prepared from commercial raw materials. Fourier-transform infrared spectroscopic measurement revealed that the degree of hydrogen bonding among hard segment chains decreased and increased with decreasing film thickness for strong and weak microphase separation systems, respectively. The microphase-separated structure, which is formed from hard segment domains and a surrounding soft segment matrix, were observed by atomic force microscopy. The size of hard segment domains decreased with decreasing film thickness, and possibility of specific orientation of the hard segment chains was exhibited for both systems. These results are due to decreasing space for the formation of the microphase-separated structure.

1. Introduction

Block copolymers are quite attractive materials to create microelectronic devices, data storage system, membranes, etc, because they form a highly ordered fine structure spontaneously.[1,2] Since different block sequence is chemically connected each other, each component can form not large domains but a nanometer sized structure. Thus, the size of the microphase-separated structure is controllable by change in the each block length. The structure itself as well as the size of microphase-separated domains can also be controlled by the ratio of block sequence, incorporation of another component (triblock copolymer) and so on. To employ a block copolymer for an application mentioned above, the films should be thin enough. Due to surface and interfacial effects, the thickness of the film can cause a different morphology than the equilibrium bulk morphology to have the lowest free energy, resulting in a changing morphology as the film thickness changes.[3]

Polyurethane (PU) can be classified into a multi-block polymer, which possesses strong polar groups in the molecular chains.[4] It is well-known that the PUs consist of the hard and soft segments, and exhibit a microphase-separated structure. In terms of thin film preparation, it is expected that the strong polar groups will help to form a stable film even at a molecular level and strongly affect the microstructure of the polymer films. Also, the multiblock structure will

[3] Author to whom all any correspondence should be addressed.

Journal of Physics: Conference Series **184** (2009) 012028 doi:10.1088/1742-6596/184/1/012028

restrict the segregation of lower surface free energy components at the film surface, resulting in the formation of the lateral phase separation. Furthermore, the strong polar groups of urethane groups are expected to show a new property, for example, ferroelectric devices.

In this paper, we investigated the effect of film thickness on the chain and microphase-separated structures of ultrathin PU films.

2. Experiments

2.1. Sample preparation

Polyurethanes were synthesized from poly(oxytetramethylene) glycol (M_n=2000), 4,4'-diphenylmethane diisocyanate (MDI) and 1,4-butandiol (BD) by a prepolymer method. This PU is strongly microphase-separated. The hard segment contents were 34 and 45 wt%. One more PU with strong microphase sepration were synthesized from poly(oxypropylene) glycol (PPG: M_n=2000), MDI and a chain extender with a controlled chain length (OU5). Another PU with weak microphase separation was synthesized from PPG, 2,4-tolyrene diisocyanate (TDI) and BD by a prepolymer method. The hard segment content was 25 wt%. Thin PU films were prepared by spin coating. The concentration of a tetrahydrofuran solution was changed to control the film thickness. Detailed descriptions of the experimental procedure are described in earlier publication from this research group.[5]

2.2. Chain and microphase-separated structures

Fourier-tranform infrared (FT-IR) spectra were collected using a Biorad FTS-3000 equipped with a mercury-cadmium-telluride detector to investigate the molecular orientation of the hard segment chains in the PU films. For the transmission FT-IR measurement, a double side polished silicon wafer was employed. p-Polarized infrared beam was used and incident angles were 10 and 74°.

A microphase-separated structure of the PU films was observed using a Seiko Instrument atomic force microscopy (AFM). Since the PU includes the rubbery soft segments at room temperature, we took an intermittent mode for the observation. Both topography and phase images were collected simultaneously. To avoid the destruction of the sample surface, the images were acquired under ambient condition using light to moderate tapping, r=0.90-0.80, where r=(set point amplitude/free amplitude of oscillation). Cantilevers employed were SI-DF-20 with spring constant of 14 N m^{-1} and resonance frequency of 130 kHz.

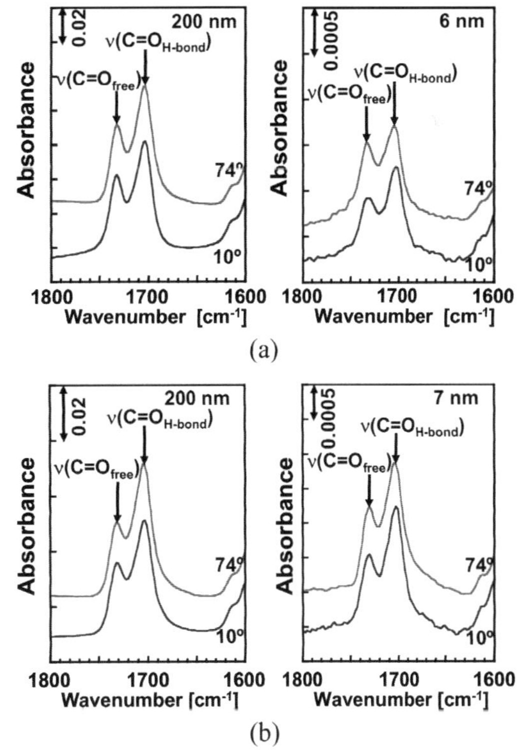

Figure 1. The FT-IR spectra for the thin PU films with (a) 34 and (b) 45 wt% of hard segment contents at incident angles of 10 and 74°.

3. Results and discussion

Figure 1 shows the FT-IR spectra for the thick and thin PTMG-MDI-BD PU films with hard segment contents of (a) 34 and (b) 45 wt% measured at incident angles of 10 and 74°. In the spectra shown here, one can see two peaks at 1704 and 1730 cm^{-1}. These peaks can be assigned to the hydrogen bonded carbonyl stretching band ($\nu(C{=}O_{H\text{-}bond})$) and free one ($\nu(C{=}O_{free})$), respectively. The intensity ratio of $\nu(C{=}O_{free})$ peak, to $\nu(C{=}O_{H\text{-}bond})$ one ($I\nu(C{=}O_{free})/I\nu(C{=}O_{H\text{-}bond})$) of the PTMG-MDI-BD-45 film are smaller than for the PTMG-MDI-BD-34 one. This is because microphase separation was progressed as the hard segment content increases. Figure 2 (a) shows thickness dependence of $I\nu(C{=}O_{free})/I\nu(C{=}O_{H\text{-}bond})$ for the PTMG-MDI-BD films with hard segment contents (HSCs) of 20, 34 and 45 wt%. $I\nu(C{=}O_{free})/I\nu(C{=}O_{H\text{-}bond})$ for the PTMG-MDI-BD-34 and -45 films increased with decreasing film thickness at around 10 nm thickness. This roughly indicates the chain structure below 10 nm thickness is quite different from that of thicker films. Discussion on a chain structure in detail will be given later. On the other hand, the magnitude of $I\nu(C{=}O_{free})/I\nu(C{=}O_{H\text{-}bond})$ for the PTMG-MDI-BD-20 film exhibited the quite large magnitude at the thicker region, indicating that the hard segment chains are dispersed into a soft segment phase. This is due simply to a phase mixing trend of the sample. And the magnitude of $I\nu(C{=}O_{free})/I\nu(C{=}O_{H\text{-}bond})$ decreased with decreasing film thickness. In other words, decreasing film thickness causes a microphase-separated trend. As the variety of $I\nu(C{=}O_{free})/I\nu(C{=}O_{H\text{-}bond})$ with the film thickness exhibited opposite trend depending on the degree of microphase separation, another two samples, which are strongly microphase-separated and microphase-mixing systems, were investigated.

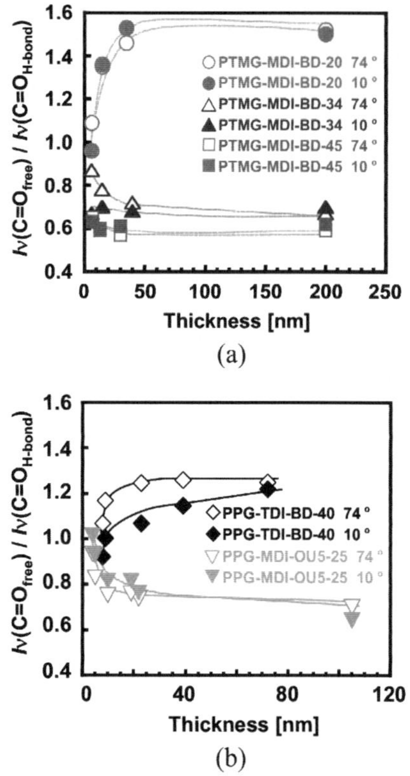

Figure 2. Thickness dependence of intensity ratio of $\nu(C{=}O_{free})$ peak, to $\nu(C{=}O_{H\text{-}bond})$ one for (a) the PTMG-MDI-BD films with hard segment contents (HSCs) of 20, 34 and 45 wt% and (b) the PPG-MDI-OU5 film with a HSC of 25 wt% and PPG-TDI-BD film with a HSC of 40 wt%.

Figure 2 (b) shows the thickness dependence of $I\nu(C{=}O_{free})/I\nu(C{=}O_{H\text{-}bond})$ for the PPG-MDI-OU5 film with a HSC of 25 wt% and PPG-TDI-BD one with a HSC of 40 wt%. For the PPG-MDI-OU5 film, which is a microphase-separated system, trend was the same with that for the PTMG-MDI-BD-34 and -45 films. On the other hand, for the PPG-TDI-BD film, which is a microphase mixing system, the trend was the same with the PTMG-MDI-BD-20 film. Furthermore, the peak position of $\nu(C{=}O_{H\text{-}bond})$ for the PPG-MDI-OU5-25 film was shifted to lower wavenumber region with decreasing film thickness.(not shown) Therefore, it seems reasonable to consider that strong constrain is induced in the ultrathin state, as the hard segment chains with strong aggregation force possess a preferable domain size in bulk.

Measurement at different incident angle in FT-IR measurement gives us important information on molecular orientation. For instance, one can estimate the direction of $\nu(C{=}O)$ in the film. It is well-known that the MDI-BD-based hard segment chains crystallize in the triclinic structure and form a plane formed with a zigzag chain.[6] The results that the magnitude of $I\nu(C{=}O_{free})/I\nu(C{=}O_{H\text{-}bond})$ measured at 74° is smaller than at 10° imply that a plane formed with a zigzag chain, in which the hard segment chains form this in the crystal, orient to the surface normal. This is the case for the PPG-MDI-OU5-25 films. However, PTMG-MDI-

BD 34 and -45 did showed opposite trend. Currently, we do not have concrete interpretation for this result. We will discuss this elsewhere.

Figure 3 shows the AFM phase images of the PTMG-MDI-BD-34 films with various film thicknesses and the PPG-TDI-BD-40 one. No characteristic feature was detected for the PPG-TDI-BD-40 film. The brighter part corresponds to the larger phase lag region. It is quite hard to interpret the phase images because these images are formed by some factors, for example, surface elasticity, adhesion force and so on. Concerning the occupied area, it seems reasonable to consider that the darker domains and a brighter matrix correspond to the hard and soft segments, respectively. As one can see the size of the microphase-separated structure seems to be different depending on the film thickness, that is, the size of domains decreased with decreasing film thickness. For the PTMG-MDI-BD-34 films, the size of microphase-separated domains was depressed at the thickness of 20 or 6 nm. The interdomain spacings of the PTMG-MDI-BD-34 films were estimated to be 27 and 21 nm for 200 and 6 nm thicknesses, respectively, by fast Fourier transform (FFT) images. Same things happened for the PTMG-MDI-BD-45 films. Furtheremore, grazing incident small-angle scattering measurement gave consistent results.[5] Therefore, we can reasonable to conclude that the microphase-separareted structure can be formed even on the condition that the thickness is much smaller than the interdomain spacing for bulk, and the interdomain spacing decreases at the ultrathin state.

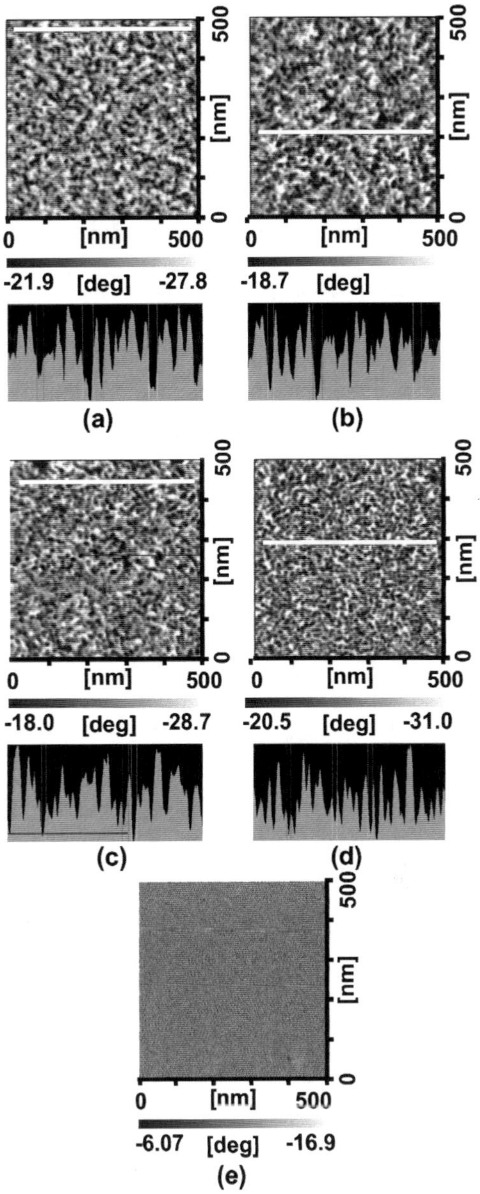

Figure 3. AFM phase images and profiles of the PTMG-MDI-BD-34 films with various film thicknesses. (a) 200, (b) 35, (c) 14 and (d) 6 nm. (e) AFM phase image of the PPG-TDI-BD 40 film.

References

[1] Fredrickson G H and Bates F S 1996 *Annu. Rev. Mater. Sci.* **26** 501
[2] Hawker C J and Russell T P 2005 *MRS Bull.* **30** 952
[3] Knoll A, Horvat A, Lyakhova K, Krausch G, Sevink G, Zvelindovsky A and Magerle R 2002 *Phys. Rev. Lett.* **89** 035501/1
[4] Ng H N, Allegrezza A E, Seymour R W and Cooper S L 1972 *Polymer* **14** 255
[5] Kojio K, Uchiba Y, Mitsui Y, Furukawa M, Matsunaga H, Sasaki S and Okda H 2007 *Macromolecules* **40** 2625
[6] Blackwell J and Lee C D 1984 *J. Polym. Sci. Polym. Phys.* **22** 759

The interface dynamics of bicontinuous phase separating structure in a polymer blend

Haruko Saito[1,*], Masahiro Yoshinaga[1], Takaaki Mihara[1], Toshio Nishi[2], and Hiroshi Jinnai[1]

[1]Department of Macromolecular Science and Engineering, Kyoto Institute of Technology, Kyoto 606-8585, Japan

[2]World Premier International, Advanced Institute for Materials Research, Tohoku University, Katahira, Sendai 980-8577, Japan

E-mail: n08a0001@kit.ac.jp

Abstract. The time evolution of the phase separating pattern during the spinodal decomposition (SD) of a specimen was observed *at the same volume of the specimen* using X-ray computerized microtomography (X-ray CT). A careful examination of time-dependent three-dimensional (3D) images revealed that bridge-like domains played an important role in the phase separation dynamics. In the course of the SD, some bridge-like domains became thicker, while the others became thinner. It was found that the pressure difference across the interface, which is quantified through the mean curvature of the interface, influences such interface dynamics.

1. Introduction

Pattern formation and interface dynamics during phase separation have long been the focus for many experimental and theoretical studies on nonlinear and non-equilibrium phenomena [1–2]. When a binary mixture is rapidly quenched from a single-phase to the spinodal region of a phase diagram by changing thermodynamic variables such as temperature or pressure it becomes thermodynamically unstable and then separates into two phases via spinodal decomposition (SD) [1–3].

A considerable number of scattering studies have been conducted on polymer blend systems to investigate pattern ordering in Fourier space by using light scattering (LS) [6–7] and small-angle neutron scattering (SANS) [8–9]. These studies have yielded the following principles of phase-separating structures during SD: (i) the evolution of concentration fluctuations in the early stage of SD can be well described by the linearized theory [10–11], and (ii) the phase separating structure grows with dynamic self-similarity in the later stages of SD [12–13].

Apart from the above mentioned scattering studies, studies were also conducted on the time evolution of the three-dimensional (3D) interface of a phase separating pattern by laser scanning confocal microscopy (LSCM). On the basis of the obtained 3D images, we can demonstrate both the dynamic self-similarity of the time evolution of the local interface during SD, and the global structure [14–16]. Further, a statistical analysis of the LSCM images was carried out to propose basic processes of the pattern evolution of the phase separating structure. However, a fundamental aspect of SD that has not yet been examined is the actual interface dynamics, which can be elucidated only by *in situ* observations. Nevertheless, such examinations have not yet been carried out because of difficulties such as the high chemical reactivity of the ultraviolet (UV) laser used in the LSCM; the UV laser causes cross-linking or degradation of polymers.

In order to examine interface dynamics, in this study, we use X-ray computerized tomography (X-ray CT) [17]. The radiation source is an X-ray with high penetrating power and low chemical reactivity with polymers. The spatial resolution of an X-ray CT has currently improved up to 1 μm. In this study, the time evolution of the interface development of a specimen during SD was observed at its original volume; an *in situ* observation was carried out; that is, an *in situ* observation was carried out using X-ray CT. The interface dynamics is discussed in the following sections.

2. Experimental

A polymer blend system of polystyrene (M_w = 200,000) and polymethylmethacrylate (PMMA) (M_n = 60,000) was used for the X-ray CT observations. In order to enhance the absorption contrast for the X-ray, 30% of the para position of styrene unit was brominated using the pre-existing procedure [18].

Poly(styrene-co-4-bromostyrene) (PS-Br) and PMMA thus obtained, were then dissolved in toluene to obtain a 5% solution and lyophilized. The volume fraction of PS-Br was 0.5. The obtained homogeneous blend was vacuum-pressed at 180°C to remove low-molecular impurities. The present blend system phase separated via the SD at 180°C. Hence, it was deduced that the onset of phase separation was at the beginning of the vacuum pressing. In order to prevent the occurrence of a volume change of the sample due to heat and contact with open air, the obtained polymer blend was cut further and embedded in an epoxy resin at 60°C. A small amount of Cu powder was dispersed in the epoxy resin as a position marker for the same-volume observation. The temperature for the epoxy embedding was well below the glass transition temperatures of the constituent polymers (107°C for PS-Br and 100°C for PMMA).

The obtained specimen was observed using X-ray CT at room temperature (nano focus ELESCAN, Nittetsu Elex Co., Ltd)). The tube voltage, tube current, increment angle, and magnification were observed to be 40 kV, 150 mA, 1200 projections/tomograph, and 50 ×, respectively. The time evolution of the phase separating pattern was observed by carrying out heat treatments repeatedly at 180°C and subsequently carrying out rapid quenching to room temperature. The phase separation time t of the obtained specimen was recorded at 180, 225, 240, 255, 270, 285, 360, 380, 385, 390, 395, 400, 405, 410, 415, 420 min.

The obtained projections were subjected to correction for the fan-type beam of X-ray CT [19] and were reconstructed using the filtered back-projection method [20]. The series of sliced images was volume-rendered by the Marching Cubes algorithm. Further, the obtained 3D images were subjected to coordinate conversion in order to align the positions with the aid of Cu powders dispersed in the epoxy resin.

3. Results and Discussion

Figure 1 shows the time evolution of the phase separating patterns of the PS-Br/PMMA blend obtained by X-ray CT observations. The white-colored areas in the figure show the PMMA phase, while the empty-looking spaces show the PS-Br phase. The interface dynamics was successfully observed in the same field of view *for the first time*. The periodic length of the phase separating pattern was observed to be 38.5 μm at t = 180 min; it further escalated to 86.3 μm at t = 420 min. The volume fractions of the PMMA phase were observed to be constant at around 0.5 within 1% errors over the observation time. Therefore, the local compositions of the polymers are assumed to attain the corresponding equilibrium values as described by the phase diagram.

Fig. 1 shows the periodicity of the interface: The elliptic interface denoted by red circles transformed into a flat interface.

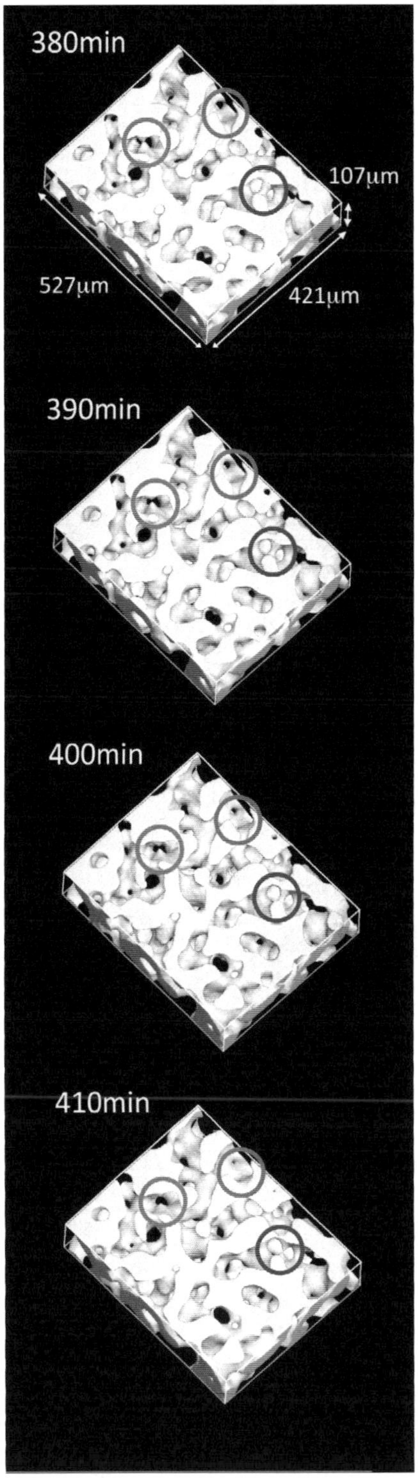

Figure 1. 3D phase separating patterns of PS-Br/PMMA blend obtained by X-ray CT observations. The white phase is the PMMA phase; the transparent phase is the PS-Br phase.

The hyperbolic interface marked by blue circles, which has bridge-like shape, was pinched off and relaxed to flatter interface. These observations are consistent with the predictions previously proposed by Jinnai *et al.* [16]. Interestingly, some bridge-like domains denoted by green circles were observed to become thicker.

While the thickening bridges concur with the overall tendency of the increase in the size of the polymers, the thinning bridge does not. On the basis of the concept that the difference between the thickening and thinning bridges should account for the intriguing behavior of SD, we calculated the curvature distributions at the interface [21]. SD exhibits a bicontinuous pattern and sustains self-similarity, though the pattern itself is not fractal architecture. The details on the curvature measurement are described elsewhere [16].

The geometry around the point of interest on a surface can be categorized by the sign of Gaussian curvature K. Surfaces with positive K, $K = 0$, and negative K are considered as elliptic, parabolic, and hyperbolic surfaces, respectively. The Gaussian curvature of the interface of phase separating pattern was plotted on the interface as depicted in Fig 2. Though slightly on the negative side, the value of K is approximately 0 on a greater part of the interface. Regardless of its thickening or thinning behavior, the bridge-like domains are identified as interfaces with negative K.

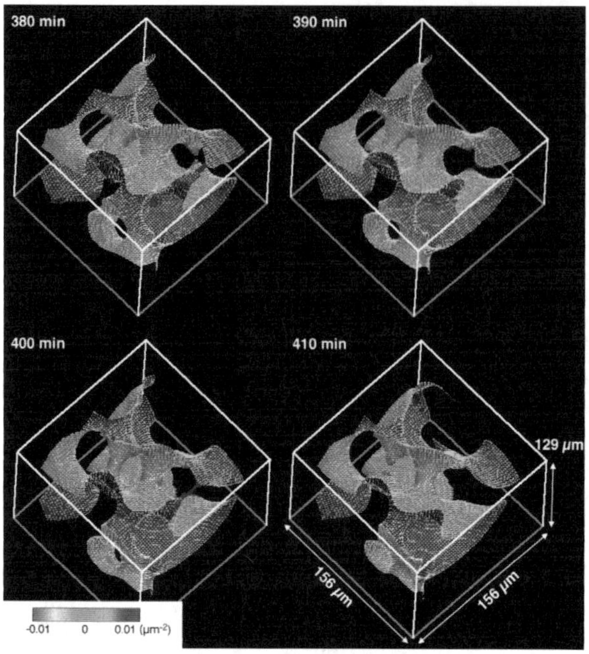

Figure 2. Time evolution of Gaussian curvature, K. Most of the interface was categorized as a hyperbolic interface; the elliptic interface shrank and disappeared.

The mean curvature appears to be negative on most of the bridge-like domains. Considering their time evolution, the thickening bridge has positive H (see also Fig. 4(a)) and the thinning bridge has negative H (see also Fig. 4(b)). Then, we can consider the Laplace equation [22],

$$\Delta P = \sigma(1/R_1 + 1/R_2) = 2\sigma H, \qquad (1)$$

where H is related to the interface pressure difference ΔP and the interfacial tension σ. Since the compositions of the coexisting phases already attain equilibrium, σ becomes a constant in the later stages of SD. Therefore, the interfacial pressure difference is proportional to the mean curvature at a given point.

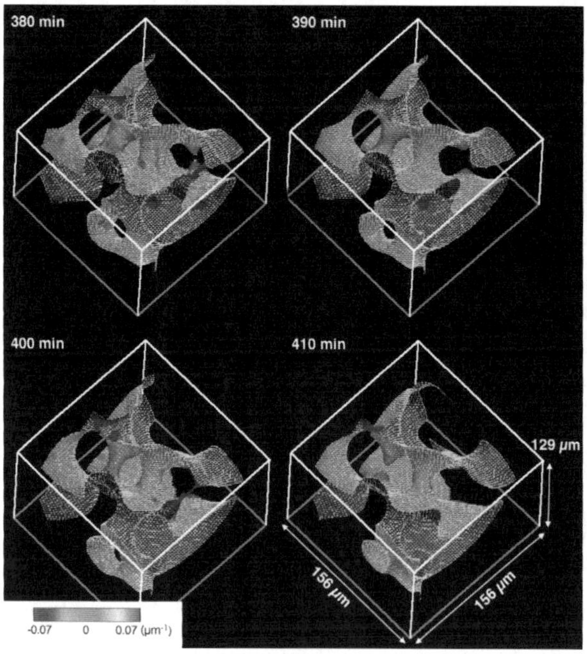

Figure 3. Time evolution of mean curvature, H. The bridge-like domain with negative H is pinched off and relaxed to a flatter interface.

On the thickening bridge, a positive mean curvature is observed; this indicates a positive interfacial pressure difference ΔP. In other words, the pressure of the PMMA phase is higher than that of the PS-Br phase. In order to relax the local pressure difference, the interface is presumed to change its shape so that there is a decrease in the positive H, resulting in the thickening of bridge-like domains. It encourages the local flow of PMMA into the bridge-like domain. On the other hand, a negative mean curvature is observed at the thinning bridges. The pressure of PS-Br phase is noted to be higher than that of the PMMA phase, and the pattern of local flow of PS-Br is observed to be opposite to that of the local flow of PMMA, i.e., the local flow is observed to be out of the bridge-like domains. Thus, by using curvature, the interfacial dynamics was found to be reasonably correlated with the interfacial geometry.

The bicontinuous structure can be regarded as the architecture of thickening and thinning bridges. The criterion for the thickening or thinning behavior of the bridges was determined to be the sign of mean curvatures. Therefore, development of the phase separating pattern can be considered to occur on the sacrifice of relatively thinner bridge-like domains with negative H. The separated bridge-like domains eventually become flatter in order to reduce interfacial energy. The procedure progresses regardless of the absolute value of the structural dimension, thus the phase separating pattern develops with self-similarity.

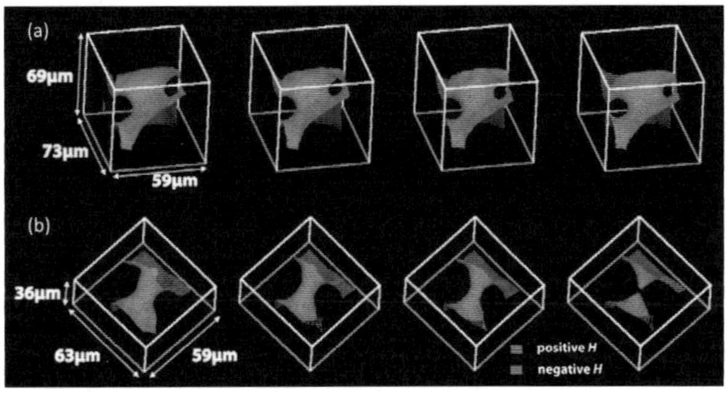

Figure 4. (a) Thickening bridge with interface having positive H and (b) thinning bridge with interface having negative H.

The interface curvature and the hydrodynamic flow steer each other. Their relation can be effectively described by the Laplace equation. However, in order to comprehend the phase separation dynamics completely, we would require a clear explanation of both interface curvature and the hydrodynamic flow. An *in situ* observation would help in the differential analysis of the interface. It would elucidate the yet ambiguous reality of the hydrodynamic flow and the physics behind the time evolution of the phase separating pattern during the later stages of SD.

4. Conclusions

Time evolution of a three-dimensional (3D) interface of phase separating pattern was observed in a mixture of poly(styrene-co-4-bromostyrene) (PS-Br) and polymethylmethacrylate (PMMA) *for the first time in the same field of view* by using X-ray computerized tomography (X-ray CT). The analysis of the interface curvature revealed that the interface dynamics can be accounted for by the pressure difference across the interface. The interface dynamics can be measured using the curvatures of the interface between the two coexisting phases.

References

[1] J. D. Gunton, M. San Miguel, P. S. Sahni, *Phase Transition and Critical Phenomena*; C. Domb, J. L. Lebowitz, Eds.; Academic Press: New York, 1983; **8** 269.
[2] T. Hashimoto, *Materials Science and Technology*; R. W. Cahn, P. Hansen, E. J. Kramer, Eds.; VCH: Weinheim 1993; **12** 251.
[3] K. Binder, *Materials Science and Technology A Comprehensive Treatment;* R. W. Cahn, P. Hansen, E. J. Kramer, Eds.; VCH: Weinheim 1990; **5** 405.
[4] T. Hashimoto. *Phase transitions*. 1988 **12** 1278.
[5] J. W. Cahn, *Chem. Phys*. 1965 **42** 93.
[6] T. Hashimoto, J. Kumaki, H. Kawai, *Macromolecules* 1983 **16** 641.
[7] H. L. Synder, P. Meakin, *J. Chem. Phys*. 1983 **79** 5588.
[8] J. S. Higgins, H. A. Fruitwala, P. E. Tomlins, *Br. Polym. J*. 1989 **21** 247.
[9] D. Schwahn, K. Hahn, J. Streib, T. Springer, *J. Chem. Phys*. 1990 **93** 8383.
[10] J. W. Cahn, *Acta Metall*. 1962 **10** 179.
[11] H. E. Cook, *Acta Metall*. 1970 **18** 297.
[12] K. Binger, D. Stauffer, *Phys. Rev. Lett*. 1974 **33** 1006.
[13] F. S. Bates, G. D. Wingnal, W. C. Koehler, *Phys. Rev. Lett*. 1985 **55** 2425.
[14] H. Jinnai, Y. Nishikawa, T. Koga, T. Hashimoto, *Macromolecules* 1995 **28** 4782.
[15] H. Jinnai, Y. Nishikawa, R. J. Spontak, S. D. Smith, D. A. Agard, T. Hashimoto, *Phys. Rev. Lett*. 2000 **84** 518.
[16] H. Jinnai, Y. Nishikawa, H. Morimoto, T. Koga, T. Hashimoto, *Langmuir* 2000 **16** 4380.
[17] M. D. Montminya, A. R. Tannenbaumb, C. W. Macosko, *J. Coll. Int. Sci*. 2004 **280** 202.
[18] B. R. Kambour, J. T. Bendler, *Macromolecules* 1986 **19** 2679.
[19] R. Ning, X. Y. Tang, D. Conover, *J. Med. Phys*. 2004 **31** 1195.
[20] D. J. DeRosier, R. A. Crowther, A. Kiug, *Proc. R. Soc*. 1970 **A317** 1.
[21] Y. Nishikawa, T. Koga, T. Hashimoto, H. Jinnai, *Langmuir* 2001 **17** 3254.
[22] A. W. Adamson, A. P. Gast, *Physical Chemistry of Surfaces*; John Wiley & Sons. Inc. New York 1997; **6** 6.

Polymer nanotechnology applied to polymeric nano-soft-materials

T Nishi[1,*], S Fujinami[1], K Nakajima[1], H Sugimori[2], M Hatta[2], M Weber[3], and H Jinnai[2,*]

[1] WPI Advanced Institute for Materials Research, Tohoku University, Sendai 980-8577, Japan
[2] Department of Macromolecular Science and Engineering, Kyoto Institute of Technology, Kyoto 606-8585, Japan.
[3] BASF AG, Polymer Research Laboratory, 67056 Ludwigshafen, Germany

E-mail: `nishi.toshio@wpi-aimr.tohoku.ac.jp`, `hjinnai@kit.ac.jp`

Abstract.
We have been developing new techniques to evaluate polymer nano-alloys and nano-composites. This "nanotechnology" can be classified into nano-three dimensional (3D) measurement, nano-physical properties evaluation systems, and nano-spectroscopy. As for the nano-3D measurement, we developed polymer oriented energy-filtered 3D transmission electron microscopy. With this method, we can access important 3D structural and elemental information that cannot be obtainable from ordinary TEM. Nano-physical properties evaluation systems were also established by developing atomic force microscopy force-distance curve measurement. The distribution of mechanical properties such as Young's modulus and adhesive energy was quantitatively obtained on high lateral resolution. These methods give new pieces of information unobtainable by the conventional techniques on polymer nanotechnology. Several results are shown here.

1. Introduction

In order to satisfy industrial demands for various purpose, vigorous research and development have been done on polymer blends and polymer composites. Recently the dispersion size of filled composite can be as small as on nanometer scale. However, the relationship between macroscopic physical properties and microscopic morphological structures was still obscure. In order to boost efficient development and promote creation of novel materials, it is important to develop techniques to evaluate nano-distribution directly. We categorized the required methods into three; nano-three dimensional (3D) measurement, nano-physical properties evaluation systems and nano-spectroscopy. We introduce the first two techniques in this paper.

Transmission electron microtomography (TEMT) is an emerging technique for observing real-space 3D images in nanoscale [1, 2]. Electron energy loss spectroscopy (EELS) is another important technology in transmission electron microscopy (TEM) that enables us to obtain element-mapped images [3]. With these two techniques together, it becomes possible to obtain elemental-mapped 3D images in materials. Such attempts have been carried out in "hard materials", e.g. metal alloys [4], but few studies on "soft materials", such as polymers, due to the severe electron beam damage to polymers. Most of the structural observations using EELS in polymer nanotechnology so far are limited to two-dimensional (2D) [5, 6]. Needless to say, such loss of one of the three dimensions results in substantial drawback in the structural observations. The growing need for 3D elemental-mapping, i.e., combined TEMT and EELS (TEMT-EELS) has been apparent.

For the purpose of evaluating local physical properties, atomic force microscopy (AFM) has a great advantage. While AFM can capture surface morphology on nanometer-scale lateral resolution, it can detect interactive force which works between a sample and a probe. AFM is mostly used in intermittent contact mode in order to observe surface morphology and phase image was thought to be responses of mechanical properties. However, to derive some physical properties from a phase image is quite difficult [7]. On the other hand, force-distance curve measurement has an advantage that it can obtain quantitative mechanical properties such as Young's modulus. A force mapping measurement is a method to measure force-distance curves at each point after dividing a sample surface into a grid. We combined force mapping measurement with force-distance curve analysis and succeeded in visualizing distribution of Young's modulus on high lateral resolution [8, 9]. We show the recent progress where adhesion-conscious theory is applied.

2. Experimental

2.1. nano-3D measurement

A poly(sulfone-*block*-amide)(PSU-*b*-PA) ($M_n = 1.5 \times 10^4$, $M_w = 5.3 \times 10^4$) was used [6]. The sulfone block contains sulfur (ϕ_{PSU}=0.54). A film specimen was prepared by casting 5 wt% hexafluoroisopropanal solutions for 2 days at room temperature and then annealed at 195 °C under vacuum for 24 h. The obtained film specimen was ultramicrotomed to the ultrathin sections (thickness: 50 nm) and stained with RuO_4 vapor for 5 minutes. TEM and TEMT images were taken using a JEOL JEM-2200FS microscope. EELS and TEMT-EELS images were taken by three-window method [3]. The electron dose has to reduce in order to suppress the electron damage during the TEMT-EELS experiments. In this experiment, relatively loose the energy resolution (±15 eV) was selected and the focus change due to the electron-energy change was determined prior to the TEMT-EELS experiments. The former allows us to use shorter exposure time and the latter to select appropriate focus without exposing excess electrons for determining the focus during the experiment. The total electron dose was 450 pA/cm^2.

2.2. Nano-physical properties evaluation

Natural rubber (NR) vulcanizates were chosen as typical industrial materials. The recipe to obtain NR vulcanizates is shown in Table 1. The vulcanizate was cured at 145°C for 30 minutes and a sheet with the thickness of 1.0 mm was made. The crosslink density was measured by swelling it in toluene and calculated with Flory-Rehner equation. The density value was 1.47×10^{-5} mol/cm^3. The average numbers of monomer units between crosslinks were 465. The NR sheet was then cut by a cryomicrotome at −180°C to expose a flat surface.

Table 1. Recipe for NR vulcanizates

NR	ZnO	stearic acid	accelerator[1]	antioxidant[2]	sulfur
100	5.0	2.0	0.5	0.5	0.5

[1] 2-mercaptobenzothiazole, [2] Santflex 13 (Monsanto).

A Nanoscope V scanning probe microscope (Veeco Metrology Group, USA) with NP-1 wide short cantilevers made of silicone nitride (Nanoprobe, USA) was employed. Nanoscope V system provides the derivation method of a spring constant of a cantilever from thermal noise, which Hutter *et al.* originally proposed [10]. The nominal value and calibrated value of spring constant was 0.58 and 0.12 N/m, respectively. The calibrated value was used in the following. We operated force mapping measurement with 50 nm of trigger deflection in air.

3. Results and Discussion

3.1. nano-3D measurement

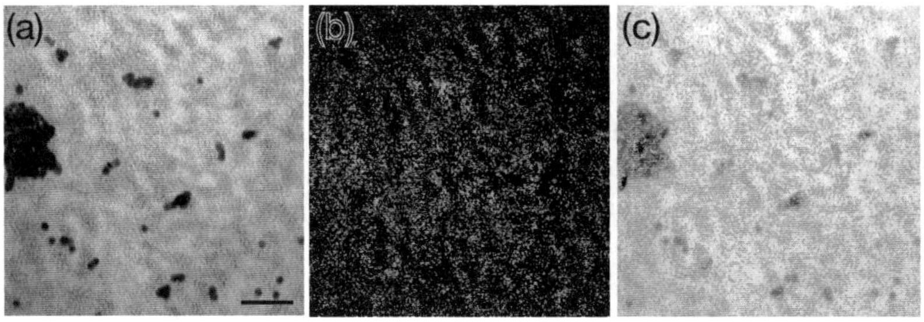

Figure 1. (a) TEM image, (b) the sulfur map obtained by three-window method and (c) superimposed image of (a) and (b) for the RuO_4-stained PSU-*b*-PA. The bar shows 50 nm.

The 3D nano-structure of the PSU-*b*-PA was investigated by TEMT-EELS. The ultrathin sections of the RuO_4-stained PSU-*b*-PA was observed by TEM (Fig. 1(a)). The gray and bright area correspond to the RuO_4-stained PSU and PA microdomains, respectively. This image suggests that PSU-*b*-PA exhibited bicontinuous-like structure, as speculated in the previous study using the same PSU-*b*-PA block copolymer [6]. The dark small dots in the image are gold nanoparticles used later as fiducial points for the alignment of the tilt series [11] and three-window method. The 2D Fourier transformation of the TEM images was carried out, from which the domain spacing of the microphase separated structure was estimated to be ca. 27.2 nm. This spacing is in agreement with the result obtained separately by small-angle X-ray scattering measurements, i.e., ca. 30 *nm* [6]. In order to obtain the sulfur map of the PSU-*b*-PA, EELS was carried out at the same area as the TEM observation (Fig. 1(a)), which is demonstrated in Fig. 1(b). The yellow area corresponds to the area with high sulfur content. The sulfur map is superimposed on the TEM image (Fig. 1(a)) and shown in Fig. 1(c). It was found that the sulfur domain exactly overlaid on the gray ones, i,e, RuO_4-stained PSU microdomains, which ensures that EELS gives precise element specific map as expected. However, since the sulfur overlapped along the depth of the ultrathin section, the structural information along this direction is lost in the 2D sulfur map. In order to grasp the 3D sulfur distribution in the specimen, it is necessary to combine EELS with TEMT (TEMT-EELS).

The resulting reconstructed 3D sulfur map of the PSU-*b*-PA obtained by TEMT-EELS is shown in Fig. 2. The yellow phase corresponds to the microdomains with high sulfur content (and hence PSU microdomains). Z-axis corresponds to the optical axis of electron microscope and also to the depth direction of the ultrathin section. The sulfur distribution is continuous in all direction in the ultrathin section: It is clear from the 3D sulfur map that the PSU-*b*-PA exhibits the bicontinuous structure. The 3D sulfur map in the PSU block copolymer was distinguished by TEMT-EELS.

3.2. Nano-physical properties evaluation

Figure 3 shows a typical force-distance curve obtained on a NR vulcanizate. Force-distance curves give the relationship between the z-scanner displacement Z and the cantilever deflection

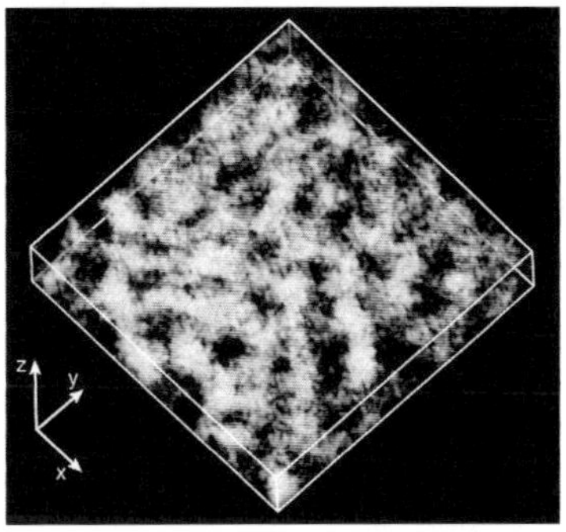

Figure 2. The 3D sulfur map obtained by TEMT-EELS of unstained PSU-*b*-PA reconstructed by the simultaneous iterative reconstruction technique (SIRT) [12]. The box size is 200 nm × 200 nm × 35 nm. The yellow phase corresponds to the the area with high sulfur content.

d. Sample deformation δ is calculated by subtracting the cantilever deflection from the scanner displacement ($\delta = Z - d$). On the other hand, applied load F is derived from the spring constant k and the deflection of the cantilever ($F = kd$). Figure 4 is force-deformation relationship derived from Figure 3 with these equations.

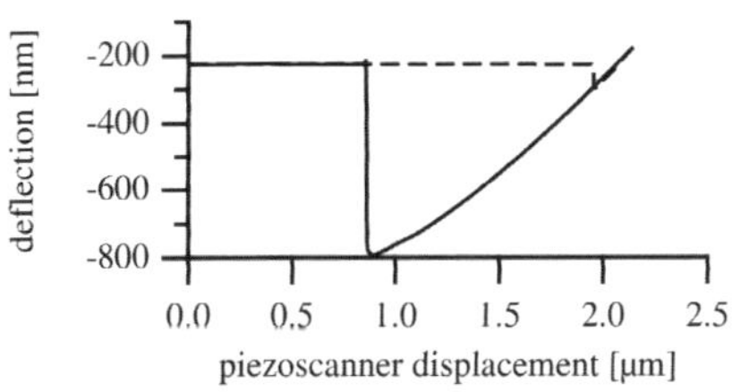

Figure 3. Typical force-distance curve on a NR vulcanizate. A dash line is on approaching, and a solid line retracting.

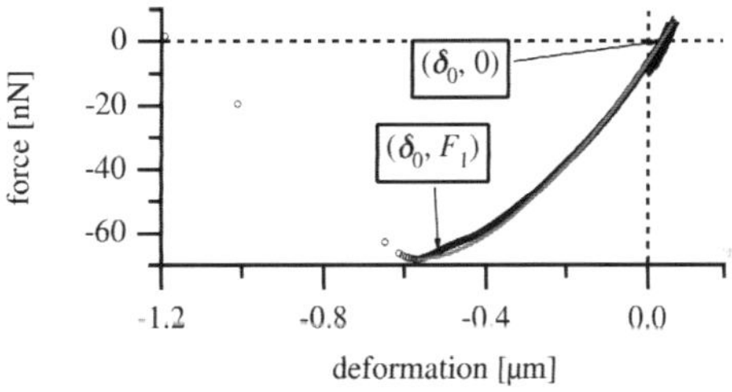

Figure 4. Force-deformation relationship derived from Figure 3. Crosshairs and circles are approaching and retracting experimental data, respectively, and a gray solid line is a calculated JKR curve.

There are several theoretical models which describe the mechanical contact between two bodies under external load [7]. Hertz theory was an old but basic theory. Derjaguin *et al.* (DMT) extended it and their limit equation of mechanical contact between an elastic plane and a solid sphere

$$F = KR^{1/2}\delta^{3/2} + F_0 \tag{1}$$

was well used for its simplicity, where K, R and F_0 are elastic coefficient, tip radius and constant adhesive force, respectively. Assuming a probe is rigid and a sample is elastic, Young's modulus E was given by the following equation

$$E = \frac{3}{4}(1 - \nu^2)K \tag{2}$$

where E and ν are Young's modulus and Poission's ratio of a sample, respectively. Our past report used it in water so as to constrain adhesion [8]. However, rubbery materials shows adhesion between cantilever, especially in air; for such condition, Hertz theory is inadequate [13].

Another theory extended by Johnson *et al.* (JKR) takes adhesion energy w into consideration [14]. The theory is described as follows.

$$F = \frac{K}{R}a^3 - 3w\pi R - \sqrt{6w\pi RF + (3w\pi R)^2} \tag{3}$$

$$\delta = \frac{a^2}{3R} + \frac{2F}{3aK} \tag{4}$$

Because these equations are unable to be converted to an explicit function of δ and F, the least mean square curve fitting is not applicable., Therefore, we applied "two-point method" proposed by Sun [15]. In this method a JKR curve was drawn so as to cross the two typical points; one point is where the attractive force and repulsive force become equivalent ($(\delta_0, 0)$ in Figure 4), and another is where adhesive force becomes maximum ((δ_1, F_1) in Figure 4). Using these two points, JKR equations are converted to the following functions where K and w are explicitly represented:

$$K = \left(\frac{1 + 16^{1/3}}{3}\right)^{3/2} \frac{F_1}{\sqrt{R(\delta_0 - \delta_1)^3}} \tag{5}$$

$$w = \frac{-2F_1}{3\pi R} \tag{6}$$

Thus, equations 2, 5 and 6 give K and w from a force-distance curve. Assuming $\nu = 0.5$ and $R = 20$ nm, the Young's modulus and adhesive energy of the curve in Figure 4 were 0.74 MPa and 0.723 J/m^2, respectively. The derived curve is depicted with a gray solid line in Figure 4. On the other hand, equation 1 based on DMT theory gave Young's modulus of 3.8 MPa. This mismatch was made by the inconsideration of adhesion in DMT theory [13]. Because of the jump-in adhesion the slope in Figure 4 became steep, thus the apparent modulus increased.

Constructing a result of a force mapping measurement onto a two-dimensional coordinate, distribution images of Young's modulus and adhesive energy were clearly visualized. The result is shown in Figure 5. Both images were simultaneously obtained from one force mapping measurement. The z-directional color scale of the adhesive energy image is inverted for comparison with Young's modulus image. Several grains whose sizes are around 1 μm were observed. Judging from an analogy in electron microscopy images, they seem to be made of ZnO. Comparing both images, obvious is a general relationship that softer area has strong adhesion. However, several deviations from the relationship were also observed. Figure 6 shows profiles of dotted line in Figure 5. While they have inverse relationship on 3–7 μm position, a hard area around 2 μm shows strong adhesion and weak adhesive area around 9 μm gives low modulus (Note that curve of adhesive energy is also inverted). It is because the origin of modulus and adhesion are not the same; adhesive energy is affected by chemical affinity but modulus not. Intermittent contact mode AFM or other method is incapable of distinguishing there differences. Force mapping measurement with JKR analysis only clarified these differences.

4. Conclusions

An example of advanced nano-3D measurement, combined TEMT and EELS (TEMT-EELS), has been given. After adjusting the energy resolution and the pre-determination of the focus change due to the electron-energy change during the EELS measurements, electron dose to the specimen was substantially reduced. The 3D sulfur map of a microphase-separated structure of a poly(sulfone-*block*-amide)(PSU-*b*-PA) was obtained using such low-dose TEMT-EELS. Bicontinuous morphology was clearly observed from the TEMT-EELS observation, which was only inferred in the previous study from the 2D TEM micrographs.

The combination of JKR analysis with force mapping measurement was shown for industrial rubber composites. Both Young's modulus distribution and adhesive energy distribution were

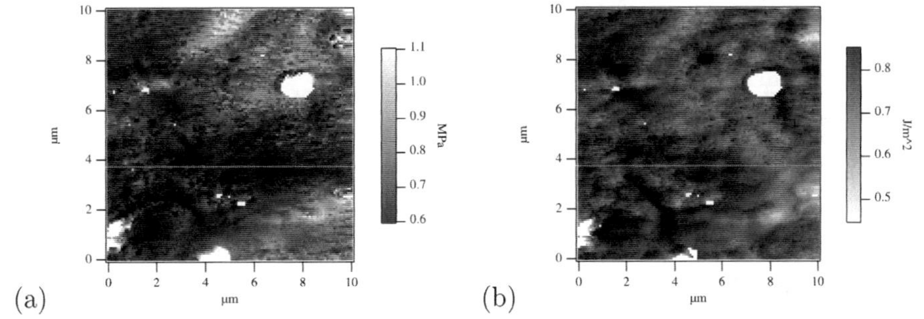

Figure 5. Result of force mapping measurement and JKR analysis. (a) Young's modulus distribution image. (b) Adhesive energy distribution image. The scan size was 10 μm and both images were obtained simultaneously. Note that the color bar of z scale in the adhesion image is inverted for comparison with the modulus image.

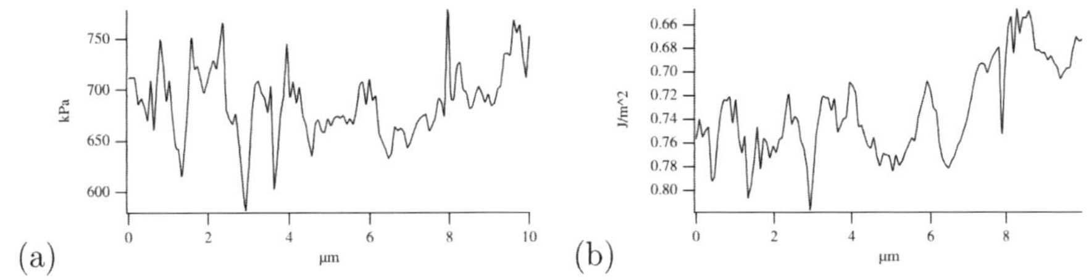

Figure 6. Section of a dotted line in Figure 5. (a) Young's modulus. (b) Adhesion energy. Note that the y scale in the adhesion section is inverted for comparison with the modulus section.

obtained with high lateral resolution. Both images roughly have inverse relationship, but not precisely. It means that material properties of different origin were separately visualized by this measurement.

These methods give new information unobtainable by the conventional techniques on polymer nanotechnology.

References

[1] Jinnai H, Nishikawa Y, Ikehara T and Nishi T 2004 *Adv. Polym. Sci.* **170** 115
[2] Jinnai H and Spontak R J 2009 *Polymer* **50** 10675
[3] Egerton R F 1996 *Electron Energy-Loss Spectroscopy in the Electron Microscope, second edition* (New York/ Plenum Press)
[4] Midgley P A and Weyland M 2003 *Ultramicroscopy* **96** 413
[5] Horiuchi S and Dohi H 2006 *Langmuir* **22** 4607
[6] Ribbe A E, Hayashi M, Weber M and Hashimoto T 2000 *Macromolecules* **33** 2786
[7] García R and Pérez R 2002 *Surf. Sci. Rep.* **47** 197
[8] Nukaga H, Fujinami S, Watabe H, Nakajima K and Nishi T 2005 *Jpn. J. Appl. Phys.* **44** 5925
[9] Nishi T, Nukaga H, Fujinami S and Nakajima K 2007 *Chin. J. Polym. Sci.* **25** 35
[10] Hutter J L and Bechhoefer J 1993 *Rev. Sci. Instrum.* **64** 1868
[11] Lawrence M C 1992 *Thee-dimensional Imaging with the Transmission Electron Microscope* ed Frank J (New York/ Plenum Press)
[12] Gilbert P 1972 *J. Theor. Biol.* **36** 105
[13] Johnson K L and Greenwood J A 1997 *J. Colloid Interface Sci.* **192** 326
[14] Johnson K L, Kendall K and Roberts A D 1971 *Proc. R. Soc. Lond.* A **324** 301
[15] Sun Y, Akhremitchev B and Walker G C 2004 *Langmuir* **20** 5837

AFM studies on microtribology of concentrated polymer brushes in solvents

Y. Tsujii, A. Nomura, K. Okayasu, W. Gao, K. Ohno, and T. Fukuda

Institute for Chemical Research, Kyoto University, Uji, Kyoto 611-0011, Japan

tsujii@scl.kyoto-u.ac.jp

Abstract. In concentrated polymer brushes (CPBs) prepared by surface-initiated living radical polymerization, graft chains in a good solvent are highly stretched, nearly to their full length, giving properties quite different and unpredictable from those of semi-dilute polymer brushes (SDPBs) previously studied. This paper reviews our recent studies on the microtribology of confronted CPBs in solvents. Atomic force microscopy revealed that the CPBs of poly(methyl methacrylate) and poly(styrene) showed dramatically low frictional coefficients in good solvents compared with those of the corresponding SDPBs. This was reasonably ascribed to confronted CPBs hardly mixing with each other due to their huge osmotic and elastic interactions. The lubrication mechanism of CPBs was discussed according to the sliding-speed dependency of frictional coefficient, and two different mechanisms, boundary and hydrodynamic lubrications, were suggested. Thermo-responsive lubrication was also demonstrated for the CPB of poly(*N*-isopropyl acrylamide). We believe that these studies will open up a new route to precision surface modification for "super" lubrication.

1. Introduction

Polymers densely end-grafted on a solid surface are obliged to stretch away from the surface, forming a so-called "polymer brush", which can dramatically change various surface properties including optical, electrical, chemical, thermodynamic, mechanical, rheological, and tribological ones.[1-5] Because of their important role in many areas of science and technology, polymer brushes have been extensively studied theoretically and experimentally. These polymer brushes were mostly categorized into the "semi-dilute" polymer brush (SDPB), in which polymer chains swollen in a solvent overlap with each other but their volume fraction is still so low that the free energy of interaction can be approximated by a binary interaction, and the elastic free energy, by that of a Gaussian chain.[6-8]

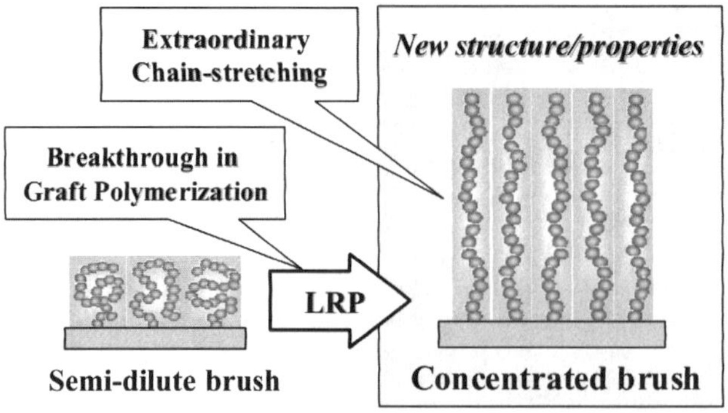

Figure 1. Conceptual illustration from SDPB to CPB.

Successful application of living radical polymerization (LRP) made it possible to graft various low-polydispersity polymers including simple homopolymers, end-functionalized polymers, block/random/gradient copolymers, and functional polymers.[9-14] At the same time, this technique brought about a striking increase of graft density: its maximum value for e.g., poly(methyl methacrylate) (PMMA) grafted on a silicon wafer reached 0.6 chains/nm^2.[15] This density was more than 1 order of magnitude higher than those of typical SDPBs, going deep into the regime of concentrated polymer brushes (CPBs)[16] which had been little explored systematically because of the unavailability of such brush samples. The CPBs of PMMA swollen in a good solvent exhibited an equilibrium film thickness as large as 80-90 % of the full (contour) length of the graft chains, indicating that the chains are extended to a similarly high degree.[17,18] Such CPBs have been found to show properties quite different and even unpredictable from those of SDPBs.[11,12] In this paper, we review our recent studies focusing on the tribological properties of CPBs in solvents.

2. Experimental section

The CPB samples were synthesized by surface-initiated atom transfer radical polymerization (ATRP), as described previously.[15] The number- and weight-average molecular weights (M_n and M_w) are the values, as determined by gel permeation chromatography (GPC), for the free polymers which were simultaneously produced from the free (unbound) initiator added to the polymerization solution for controlling graft polymerization. There are reasons to believe that the molecular characteristics of the free polymers closely approximate those of the graft polymers.[19-22] Table 1 lists the molecular characteristics of the CPB samples used in this study. The SDPB samples were prepared by the so-called "grafting-to" method, for which a polymer with a triethoxysilyl group at the chain end was synthesized by the ATRP using (2-bromo-2-methyl)propionyloxypropyltriethoxysilane (BPE) as a free initiator. The layer thickness L_d, in a dry state, of the brushes on a silicon wafer was determined by spectroscopic ellipsometry (M-2000U, J. A. Woolam, USA), and the graft density σ was evaluated from L_d and M_n values. It was reasonably assumed that nearly the same brushes were also formed on the silica probe particle.

The structure and compressibility of swollen brushes on a silicon wafer were measured by atomic force microscopy (AFM; NanoWizard, JPK Instruments AG, Germany or SPA-400+SPI3800N, SII NanoTechnology Inc., Japan) using an unmodified silica probe particle which was glued on a rectangular cantilever (see Figure 2). The force curves were obtained by measuring the normal force F_N (= k_N Δz, where k_N is the normal spring constant and Δz is the normal displacement of the

Table 1. Typical molecular characteristics of CPBs of poly(methyl methacrylate) (PMMA), poly(styrene) (PS), and poly(N-isopropylacrylamide) (PNIPAM) used in this study

Sample	M_n [a)]	M_w/M_n [a)]	σ / chains nm^{-2}	σ^* [b)]
PMMA	122,000	1.39	0.51	0.37
	88,000	1.17	0.53	0.38
PS	37,000	1.27	0.49	0.37
PNIPAM	50,000	1.15	0.70	0.53

a) the values for the free polymers.
b) the reduced graft density (per cross-sectional area of chain, which is estimated from the bulk density of monomer in this case); the maximum value of σ^* is unity, corresponding to the close packing of graft chain with all-trans conformation.

cantilever) as a function of separation. The measured force F_N can be reduced to the free energy G_f of interactions between two parallel plates according to the Derjaguin approximation, $F_N/R = 2\pi G_f$, where R is the radius of the probe particle.

The frictional properties between swollen brushes were also measured by AFM in the same configuration except that the silica probe particle was also modified with polymer brushes (see Figure 2). The lateral (frictional) force F_L was evaluated according to the equation, $F_L = k_L\ \theta_L/2R$, where k_L is the torsional spring constant and θ_L is the angular (torsional) displacement of the cantilever. The θ_L was monitored as a function of F_N, while the sample substrate was loaded with the silica probe particle and slided back and forth along the direction normal to the cantilever at differing sliding speeds v.

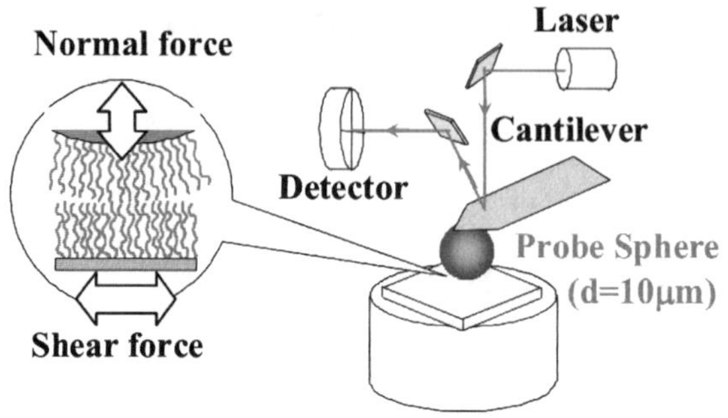

Figure 2. AFM setup for microtribology measurements

3. Results and discussion

3.1. Tribological properties of CPB in good solvent[23]

Figure 3 shows the normal force curves (without lateral modulation) for the CPB and SDPB of PMMA compressed by the (unmodified) silica probe particle in toluene (TOL). Upon compression, the F_N between the swollen brushes was completely repulsive. The data for the SDPB was nearly consistent to the scaling theory derived by de Gennes using the scaling approach.[24] The CPB ($\sigma = 0.51$ chains/nm^2, $M_n = 122{,}000$, $M_w/M_n = 1.39$), however, gave much steeper force profile, meaning that the brush layer is more resistant to compression: with $F/R > 1$ mN/m, the CPB layer was compressible only to $D/L_e \approx 0.7$, in contrast to $D/L_e < 0.2$ for the SDPB. This strong resistance against compression must be characteristic of CPB due to an extremely high osmotic pressure, being one of important properties for the development of tribomaterials.

The strong resistance against compression and dense anchoring of graft chains in CPBs are expected to drastically improve tribological properties such as friction, wear, and lubrication as compared with those of SDPBs. Figure 4 plots the frictional coefficient μ (= F_L/F_N) vs. normal force F_N (load) between confronted brushes at a sliding speed of $v = 20$ μm/s, which is in a regime of the boundary lubrication discussed below. The SDPB ($\sigma = 0.024$ chains/nm^2, $M_n = 90{,}000$, $M_w/M_n = 1.27$) had two different regime of friction: at low applied loads, the F_L and hence the μ value is very low (<0.001), and then at an onset of the threshold, the F_L steeply increases with increasing F_N, approaching a constant μ value of about 0.1 (for higher loads). Such a transition from low to high-frictional stages was already reported by Klein et. al.,[4] who measured the frictional properties between the SDPB of poly(styrene) (PS) in TOL using the

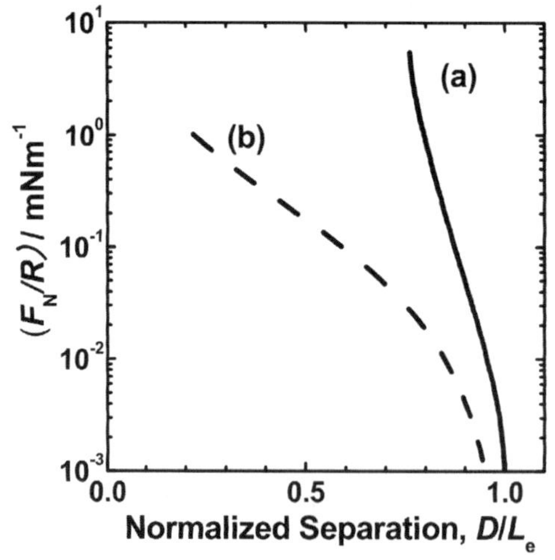

Figure 3. Plot of F_N/R vs normalized separation D/L_e for (a) CPB and (b) SDPB of PMMA in TOL.

surface force apparatus. This transition was ascribed to the interpenetration (mixing) of the brushes at high loads. More interestingly, the F_L between the CPBs of PMMA (σ = 0.53 chains/nm^2, M_n = 88,000, M_w/M_n = 1.17) showed no such transition, staying at low values, lower than 5 x 10^{-4} in the whole range of loads studied. This μ value is believed to be one of the lowest of all materials and comparable to that achieved for the SDPBs of polyelectrolyte with the help of a charge effect. This extremely low-friction property was reasonably ascribed to the fact that swollen CPBs hardly interpenetrate each other due to the large osmotic pressure and highly stretched chain conformation (entropic interactions).

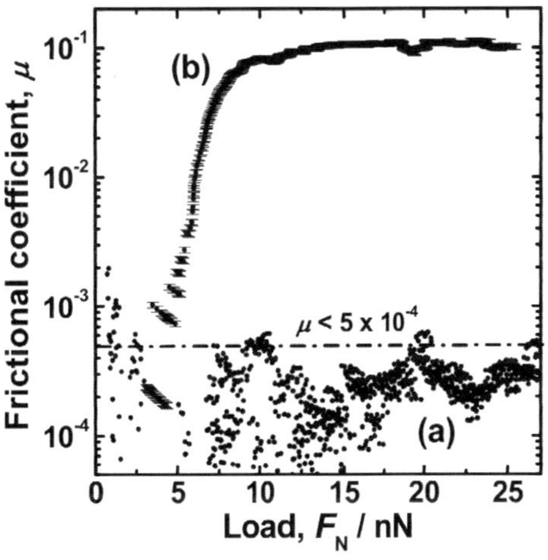

Figure 4. Plot of frictional coefficient μ vs normal load F_N for (a) CPB and (b) SDPB of PMMA in TOL.

This mechanism of "super lubrication" of CPBs should be effective not only in a microscopic scale as verified by AFM but also in a macroscopic scale. Kobayashi et al.[25,26] studied the tribological property of the CPBs by sliding the stainless or glass probe with R = 5 mm on the brush surface in the air as well as in solvents. Although the μ did not reach the low value obtained in a microscopic scale, the CPBs of PMMA were found to have better frictional property and better wear resistance than the corresponding spin-coated PMMA film. These properties may be improved by increasing the brush thickness and/or decreasing the surface roughness of substrates.

3.2. Lubrication in mixtures of good and poor solvents[27]

In order to discuss the mechanism of lubrication in detail, we measured the μ value as a function of sliding speed v for the CPB of PS (σ = 0.49 chains/nm^2, M_n = 37,000, M_w/M_n = 1.27) in mixtures of TOL and isopropanol (IPA) (as good and poor solvents, respectively). Figure 5 shows the results at F_N = 20 nN, corresponding to the high-load regime as mentioned above. It should be noted that the SDPB gave a nearly constant and high μ value independent of v (the data are not shown in the figure). First of all, we discuss the v dependency of μ in pure TOL. The data suggest the two regions in the v dependency; at a high v region, the μ value increases with increasing v. We ascribed the high v region to the hydrodynamic lubrication, in which the frictional property must be related to the viscosity of solvent (in the swollen brush). The slope of the line in this regime was estimated to be ca. 0.7. It should be unity for Newtonian fluids. The lower v region in which μ is nearly constant is ascribed to the boundary lubrication, where the interaction between brushes is important. The CPB of PMMA also gave similar v dependency of μ, and the data at v = 20 μm/s discussed in the previous section were in a regime of boundary lubrication.

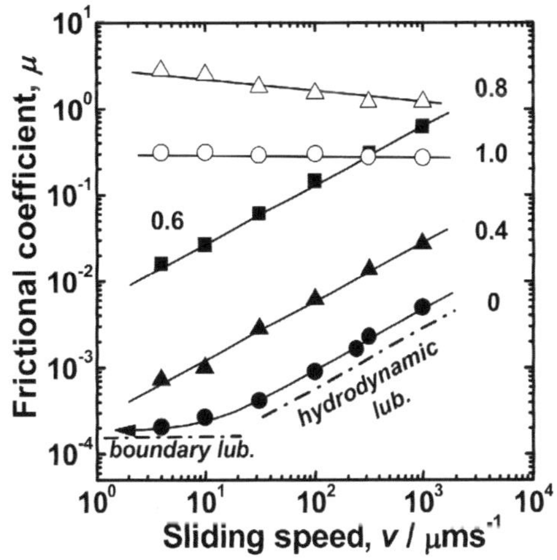

Figure 5. Plot of frictional coefficient μ vs sliding speed v for the CPB of PS in IPA/TOL mixtures; the figures indicate the IPA content.

Figure 5 suggests that with increasing IPA content up to at least 0.6, the v dependency line corresponding to the hydrodynamic lubrication shifted to a higher value with nearly the same slope. We determined the equilibrium thickness L_e of swollen brush in these mixed solvents, and interestingly, we found that these data in the hydrodynamic lubrication could be scaled by the degree of swelling, giving $\mu \sim v^{\alpha}(L_e/L_d)^{\beta}$, where $\alpha = 0.7$ and $\beta = -4$. This would give good information to discuss the details about the hydrodynamic lubrication, which is going on by simulational and theoretical studies. Further increase in IPA content resulted in almost constant μ value independent of v. Under such poor-solvent condition, the CPB was in a non-swollen state and hence no more worked as a hydrodynamic lubricant. This might be similar to the solid-solid friction with direct contact of surfaces.

3.3. Thermoresponsive lubrication[28]

Controlled synthesis of well-defined CPBs of PNIPAM still remained challenging. For achieving this, the condition of surface-initiated ATRP was optimized; the catalytic system of CuCl/$CuCl_2$/tris[2-(dimethylamino)ethyl]amine along with a surface-initiator, (2-chloropropionyloxy)hexyltriethoxy-silane enabled a controlled growth of PNIPAM brushes in DMSO at 25 °C. The brush thickness in dry state (L_d) increased proportionally with the M_n of the (low-polydispersity) free polymer (as a good measure of the graft polymer), giving a graft density of ca. 0.7 chains/nm^2 which was expected to be deep in the CPB regime.[16] We confirmed that this CPB had a large swollen thickness comparable to the full length of graft chains and a μ value as low as 10^{-4} in ethanol, which is a good solvent for PNIPAM. Figure 6 shows (a) the degree of stretching $L_e/L_{c,w}$ of the CPB and the frictional coefficient (μ) between the CPBs in water as a function of temperature, where $L_{c,w}$ is the weight-average full length of the graft chain (in all-trans conformation). The CPB of PNIPAM ($\sigma = 0.7$ chains/nm^2, M_n = 50,000, $M_w/M_n = 1.15$) was highly swollen, nearly to their full lengths $L_{c,w}$, in water at 5 °C, gradually shrinking with increasing temperature down to an almost non-swollen state above 30 °C. This is reasonably understood in terms of the decreasing solvent quality of water for PNIPAM and hence the decreasing osmotic pressure; the CPB is more difficult to be swollen than the SDPB because of the much larger penalty of conformational entropy in the former. Interestingly, a dramatic change in μ between brushes (more than 3 orders of magnitudes) was observed at around 30 °C, below which it stayed at very low values (< 10^{-3}). This is ascribed to a dramatic change in structure/properties of the outermost surface.

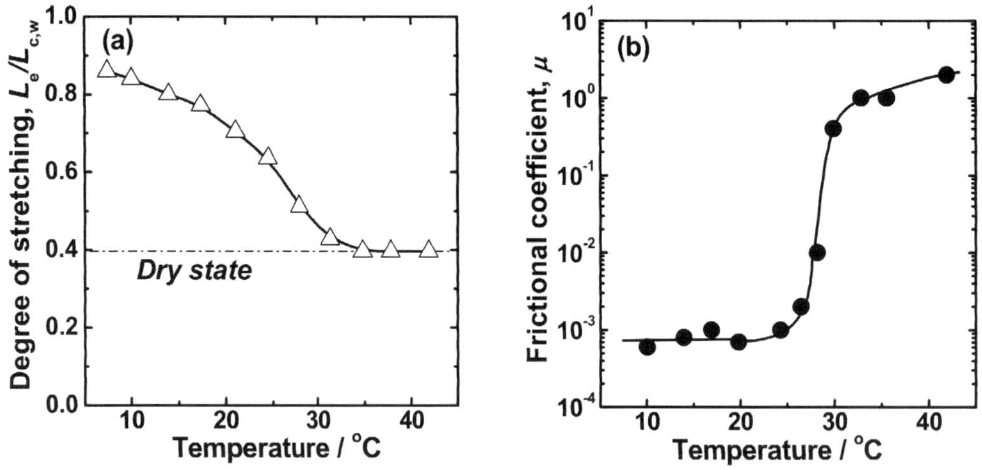

Figure 6. Plots of (a) degree of stretching $L_e/L_{c,w}$ and (b) frictional coefficient μ as a function of temperature for the CPB of PNIPAM in water.

4. Conclusion

Recent successful application of living radical polymerizations has made it possible to modify solid surfaces with various end-grafted polymers of exceptionally high graft density, allowing us to investigate deeply into the CPB regime of graft density. Among other fascinating features of CPBs, their solvent-induced super lubrication was described in this paper. In contrast to SDPBs which gave a transition in frictional coefficient with increasing applied load, CPBs gave no such transition. This unique feature of CPBs was discussed to originate in entropic interactions in common for all kinds of CPBs in good solvents. In this regard, surface-initiated LRP and the CPBs thereby obtained are opening up a versatile route to the creation of new tribomaterials that have never been realized by the conventional techniques of surface modifications. Hopefully, this progress will lead to a wealth of applications.

5. Acknowledgements

This work was supported in part by a Grant-in-Aid for Specially Promoted Research (17002007) from the Ministry of Education, Culture, Sports, Science, and Technology, Japan.

References

[1] Napper D H 1983 *Polymeric Stabilization of Colloidal Dispersions* (London: Academic Press).
[2] Raphael E and Degennes P G 1992 *J. Phys. Chem.* **96** 4002.
[3] Klein J 1996 *Annu. Rev. Mat. Sci.* **26** 581
[4] Klein J and Kumacheva E 1995 *Science* **269** 816.
[5] Parnas R S and Cohen Y 1994 *Rheol. Acta* **33** 485.
[6] Israelachvili J N 1992 *Intermolecular and Surface Forces 2nd ed* (London: Academic Press).
[7] Halperin A, Tirrell M and Lodge T P 1992 *Adv. Polymer Sci.* **100** 31-71.
[8] Kawaguchi M and Takahashi A 1992 *Adv. Colloid Interface Sci.* **37** 219.
[9] Edmondson S, Osborne V L and Huck W T S 2004 *Chem. Soc. Rev.* **33** 14.
[10] Advincula R C, Brittain W J, Caster K C and Ruehe J (Editors) 2004 *Polymer Brushes* (Weinheim: Wiley-VCH).
[11] Tsujii Y, Ohno K, Yamamnoto S, Goto A and Fukuda T 2006 *Adv. Polymer Sci.* **197** 1
[12] Fukuda T, Tsujii Y and Ohno K 2007 *Macromolecular Engineering. Precise Synthesis Materials Properties Applications*, ed Matjaszewski K, Gnanou Y and Leibler L (Wiley-VCH) Chapter 11.
[13] Pyun J, Kowalewski T and Matyjaszewski K 2003 *Macromol. Rapid Commun.* **24** 1043.
[14] Hawker C J, Bosman A W and Harth E 2001 *Chem. Rev.* **101** 3661.
[15] Ejaz M. Yamamoto S. Ohno K. Tsujii Y. Fukuda T 1998 *Macromolecules* **31** 5934
[16] Ohno K. Morinaga T. Takeno S. Tsujii Y. Fukuda T 2007 *Macromolecules* **40** 9143
[17] Yamamoto S, Ejaz M, Tsujii Y, Matsumoto M and Fukuda T 2000 *Macromolecules* **33** 5602.
[18] Yamamoto S, Ejaz M, Tsujii Y and Fukuda T 2000 *Macromolecules* **33** 5608.
[19] von Werne T and Patten T E 1999 *J. Am. Chem. Soc.* **121** 7409.
[20] von Werne T and Patten T E 2001 *J. Am. Chem. Soc.* **123** 7497.
[21] Pyun J, Jia S J, Kowalewski T, Patterson G D and Matyjaszewski K 2003 *Macromolecules* **36** 5094.
[22] Blomberg S, Ostberg S, Harth E, Bosman A W, Horn B Van and Hawker C J 2002 *J. Polymer Sci. Part A. Polymer Chem.* **40** 1309.
[23] Okayasu K et al. to be published.
[24] de Gennes P G 1987 *Adv. Colloid Interface Sci.* **27** 189.
[25] Sakata H, Kobayashi M, Otsuka H and Takahara A. 2005 *Polym. J.* **37** 767.
[26] Kobayashi M and Takahara A. 2005 *Chem. Lett. (Jpn.)* **34** 1582.
[27] Nomura A et al. to be published.
[28] Gao W et al. to be published.

Nanoscopic Adhesion Test for Polymers by A Light-lever System

Ifu Narayama, Keiichi Akabori, Hiroshi Morita, Toshihiko Nagamura and Keiji Tanaka*

Department of Applied Chemistry, Kyushu University, 744 Motooka, Nishi-ku, Fukuoka 819-0395, Japan

E-mail: k-tanaka@cstf.kyushu-u.ac.jp

Abstract. Polymer nano-adhesion between a cantilever tip coated with polymer and a flat polymer film was studied by a light-lever system using an atomic force microscope. The polymer interface was adhered at a temperature above the surface glass transition temperature for a given time. Nano-adhesion force (F_{nano}), at which the tip was detached from the surface, was estimated from the deflection of the lever with a known spring constant. Nano-adhesion strength (G_N) was simply obtained dividing F_{nano} by the contact area, which was estimated on the basis of Johnson-Kendall-Roberts theory. The time evolution of the interfacial thickness was independently examined by dynamic secondary ion mass spectrometry. Interestingly, G_N increased with increasing interfacial thickness. However, it can be hardly judged whether G_N is proportional to the interfacial thickness with the exponent of 1 or 2. Then, temperature dependence of G_N was examined. Above the bulk glass transition temperature, the relation between temperature and G_N was well expressed by a Williams-Landel-Ferry type equation. This means that the nano-adhesion strength is governed by friction between segments. Once this is accepted, G_N should be proportional to the interfacial thickness with the exponent of 2.

1. Introduction

Recently, various kinds of organic thin-film devices such as solar cells and electro luminescence devices have been widely studied due to their flexibility, shock-resistant properties and processibility [1]. Buried interfaces between organic layers in these devices play crucial roles in their performances. Thus, it should be discussed how the interfaces are formed as well as how their structure and physical properties differ from those in the internal state. Besides, the other research trend for organic thin-film devices is a size effect [2]. The device performance might be a function of its size; the smaller would be the better because macroscopic and microscopic defects should be reduced with decreasing size. In that sense, to design and construct ultimate-functionalized devices, their size should be as tiny as possible based on either a top-down or bottom up method.

When the size of materials becomes smaller and smaller, usual methods to assemble pieces of materials, such as glue, adhesives, and so forth, cannot be used. This is because it's technically difficult to paste such viscous liquids homogeneously onto nano-spaces. So far, segmental dynamics at polymer surfaces has been widely studied by many researchers using a wide variety of experimental techniques. One of the main conclusions emerged thus far is that the surface glass-transition temperature (T_g^s) is lower than the corresponding bulk value [3]. In other words, the chain mobility at the surface is more enhanced than that in the internal phase. This leads to a demonstration that (polystyrene (PS)/PS) bilayer interface can be evolved in terms of the thickness even at a temperature

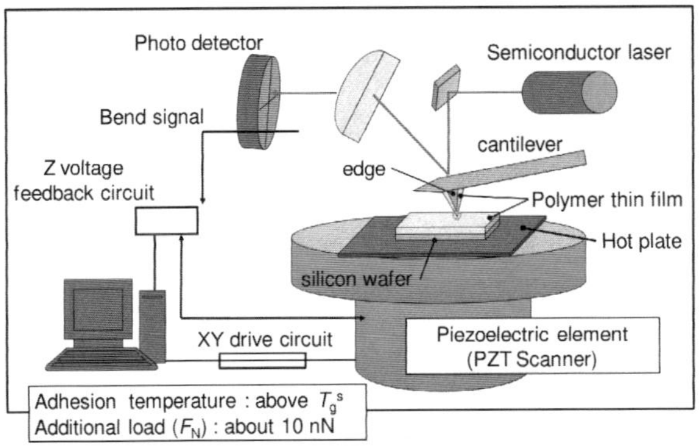

Figure 1. Schematic illustration of nano-adhesion measurement with an AFM.

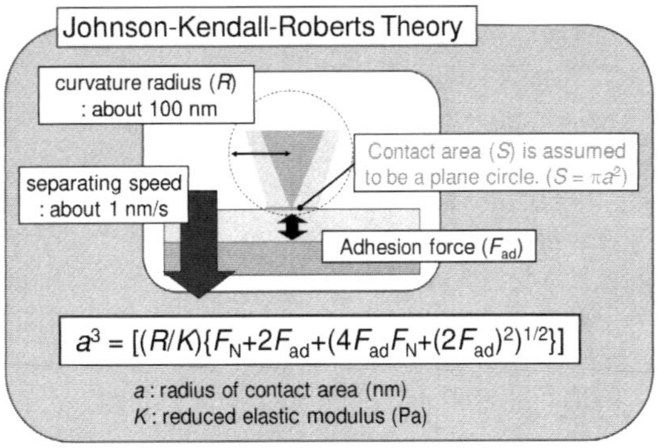

Figure 2. Contact model between a tip and a film based on Johnson-Kendall-Roberts (JKR) theory.

being well below the bulk glass-transition temperatures (T_g^b). Then, this knowledge was applied to adhesion between PS films with macroscopic area. Even if the adhesion temperature (T_{ad}) was lower than the T_g^b, adhesion strength for the PS bilayers increased with annealing time on account of the segmental interdiffusion within 10 nanometers [4]. This behavior could be simply explained in terms of the enhanced mobility for chains in the interfacial region, which was originally the film surface. Based on these results, it seems reasonable to infer that tiny peaces of polymer materials can be assembled for one another without using usual adhesives if only the surface mobility is activated. However, the lateral area should be adhered was macroscopic for the previous experiment. Hence, it is necessary to discuss what happens with the adhesion if the adhered area also decreased to be nanometer scale. The objective of this study is to give a direct answer to this question, nano-adhesion measurement.

2. Experimental

As a polymer, monodisperse PS (M_n = 26,000, M_w/M_n = 1.06) was used. The T_g^b value for PS was evaluated to be 372 K by differential scanning calorimetry (DSC). Figure 1 shows the schematic illustration for nano-adhesion measurement conducted by a light-lever system of an atomic force microscope (AFM). A tip was gently dipped into a dilute solution of PS with the aid of a micrometer and then took out from the solution right away. Then, the tip was dried and well annealed to relax the preparation history. Also, a film of PS was spin-coated onto silicon wafers and was also annealed to reach a quasi-equilibrium state. The tip coated with PS was attached to the film under a gentle pressure in the AFM vacuum chamber. The interface was allowed to be developed at a temperature above the T_g^s. After a given time (t_{ad}), nano-adhesion force (F_{nano}), at which the tip was detached from the film surface, was evaluated on the basis of the deflection of the lever with a known spring constant. Nano-adhesion strength (G_N) was finally obtained dividing F_{nano} by the contact area (πa^2) where a is radius of contact area.

$$G_N = F_{nano} / \pi a^2 \quad (1)$$

Contact area made upon the adhesion process is not always the same, depending on original tip shape, adhesion time, temperature, and so forth. Hence, the adhered area for each run was calculated on the basis of a traditional contact mechanics proposed by Johnson-Kendall-Roberts [5], as shown in Figure 2. According to this framework, the a value is given by

$$a^3 = [(R/K)\{F_N + 2F_{ad} + (4F_{ad}F_N + (2F_{ad})^2)^{1/2}\}] \quad (2)$$

where R, K, F_N and F_{ad} are radius of the curvature for the tip, reduced elastic modulus, normal force applied and adhesion force, respectively.

Journal of Physics: Conference Series **184** (2009) 012032 doi:10.1088/1742-6596/184/1/012032

3. Results and Discussion

3.1. Time evolution of nano-adhesion strength. Figure 3 shows the time evolution of G_N for (PS/PS) interfaces adhered at 365 K. The T_g^s for the PS used was 303 K [3]. That is, the adhesion temperature (T_{ad}) was between the values of T_g^s and T_g^b. When t_{ad} was shorter than 10^3 s, the G_N remained almost constant. This time independence implies that the adhesion strength in this region may be generated by a static factor such as van der Waals force rather than a dynamic one. Once t_{ad} went beyond 10^3 s, the G_N first increased with increasing t_{ad}, and then, reached a constant value of about 5 MPa after 10^5 s. The solid and dotted lines, which are both the guide of the eye, possess the slopes of 1/2 and 1/4 to the G_N-t_{ad} relation. The physical meaning of this exponent is later discussed.

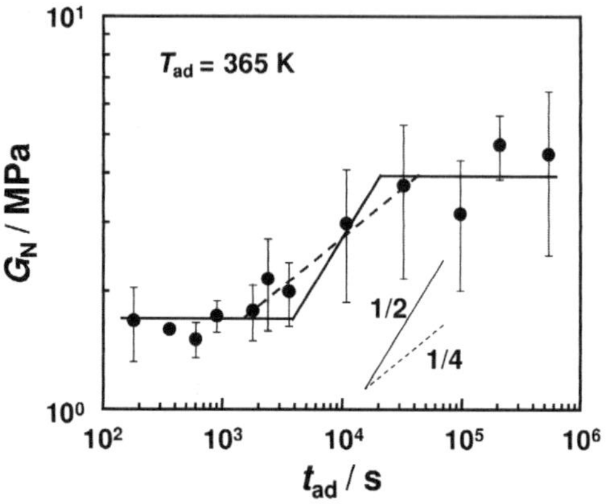

Figure 3. Time evolution of G_N for (PS/PS) interfaces adhered at 365 K.

We also examined the time evolution of interfacial thickness (d) at 365 K by dynamic-secondary ion mass spectrometry (DSIMS) [6]. In this case, the interface became thicker with increasing time until a given time, and then reached a constant. As segments reach in the depth region in which relatively large scale molecular motion, such as segmental motion, is frozen, the interdiffusion ceases. The slope for the t_{ad} vs. d relation was 1/4. According to a dynamic picture of polymer chains by Doi and Edwards [7], this exponent indicates that not chains but segments interdiffused across to the interface. Since the interface is originally formed by two surfaces, a half of the interface corresponds to a region where segments can be moved. The half thickness of the interface being in a quasi-equilibrium state was evaluated to be approximately 4 nm, as shown in Figure 3, and was smaller than the spatial dimension of a PS chain used here. Taking into account these, it seems reasonable to claim that the adhesion strength is generated by the interdiffusion of segments.

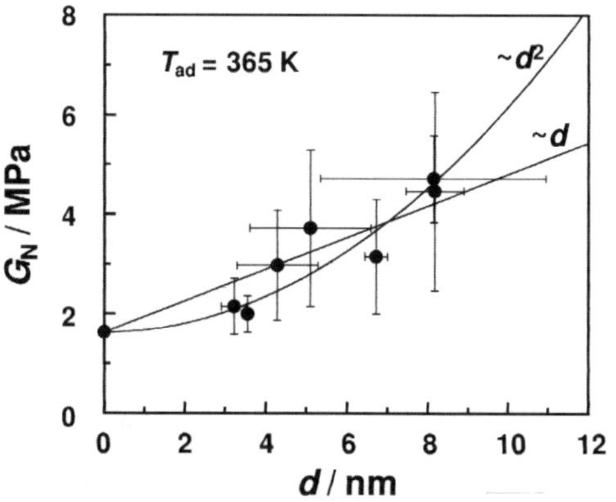

Figure 4. Relationship between d and G_N for (PS/PS) interfaces adhered at 365 K.

3.2. Relation of nano-adhesion strength to interfacial thickness. We now come to a direct comparison between d and G_N at a given t_{ad}. Figure 4 shows the result for the (PS/PS) interface adhered at 365 K. When the t_{ad} was shorter than 10^3 s, the interface could not be experimentally obtained because the interfacial broadness was comparable to the detection limit of DSIMS. Thus, in this t_{ad} region, the d value was assumed to be 0. The G_N value simply increased with d. The solid and dotted curves are drawn for the guide of the eye. The former and latter denote the relations of $G_N \sim d^2$ and $\sim d$, respectively.

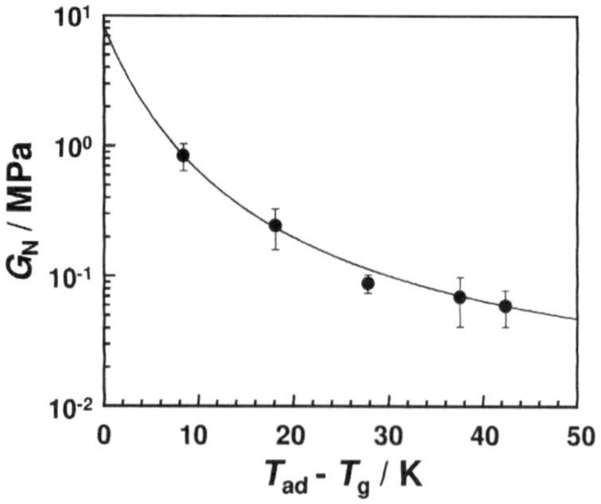

Figure 5. T_{ad} dependence of G_N for (PS/PS) interfaces adhered at temperatures above T_g^b.

While the relation of $G_N \sim d$ seems to be reasonable for glassy polymers [5], this may not be the case for rubbery chains. Schnell and co-workers have proposed that adhesion strength between rubbery polymers results from the resistance force based on monomer friction [8]. In this model, so-called chain pull-out model, G_N is expressed by the following,

$$G_N = \sum f_{mono} d^2 (2C_\infty l_0^2)^{-1} \propto A f_{mono} d^2 \tag{3}$$

where Σ, f_{mono}, C_∞, l_0 and A are areal density of chains, monomeric friction coefficient, characteristic ratio and C-C bond length, constant, respectively. That is, the G_N value is proportional to the square of d. It can be hardly judged only from Figure 4 whether which model is appropriate to explain our current experimental results.

3.3. Temperature dependence of nano-adhesion strength. According to the chain pull-out model, the G_N value for the (PS/PS) interfaces being in a quasi-equilibrium state should be dominated by the monomeric friction. Since the friction coefficient is a function of temperature, we next examined the T_{ad} dependence of G_N. In this case, t_{ad} was long enough to reach a constant in G_N to reach in a quasi-equilibrium state. Figure 5 shows T_{ad} dependence of G_N for (PS/PS) interfaces adhered at temperatures above T_g^b. The abscissa is expressed by (T_{ad}-T_g^b). The G_N value for the (PS/PS) interfaces decreased with increasing T_{ad}. Lodge and co-workers have reported the temperature dependence of f_{mono} for monodisperse PS, which follows the Williams-Landel-Ferry (WLF) equation above T_g^b [9].

$$\log(f_{mono} / f_{ref}) = -C_1 (T - T_{ref}) / \{C_2 + (T - T_{ref})\} \tag{4}$$

where C_x (x = 1, 2) is fitting parameters, and f_{ref} is monomeric friction coefficient at reference temperature (T_{ref}). The experimental data was well fitted by eq. (4) with C_1 of 0.866 and C_2 of 59.5, as shown in Figure 5. This makes it clear that the G_N value under the current condition is controlled by d^2. Once we accept this, the slope for the G_N-t_{ad} relation shown in Figure 3 is plausible to be 1/2 rather than 1/4. More conclusive study will be reported in the near future.

4. Conclusions

Polymer nano-adhesion measurement was successfully carried out using an AFM. The adhesion strength for the (PS/PS) interfaces could be apparently expressed both by the interfacial width and its square. The T_{ad} dependence of equilibrium G_N followed the WLF type equation. Hence it can be concluded that the adhesion strength would be dominated by the frictional force between segments.

References

[1] N. Sariciftci, L. Smilowitz, A. Heeger, and F. Wudl, *Science*, **258**, 1474 (1992).
[2] H. Hoppe, T. Glatzel, M. Niggemann, W. Schwinger, F. Schaeffler, A. Hinsch, M. Ch. Lux-Steiner, and N. S. Sariciftci, *Thin Solid Films,* **511**, 587 (2006).
[3] K. Tanaka, A. Takahara, and T. Kajiyama, *Macromolecules*, **33**, 7588 (2000).
[4] K. Akabori, D. Baba, K. Kouguchi, K. Tanaka, and T. Nagamura, *J. Polym. Sci. Part B*, **44**, 3598 (2006).
[5] K. L. Johnson, K. Kendall, and A. D. Roberts, *Proc. R. Soc. Lond. Ser. A*, **324**, 301 (1971).
[6] D. Kawaguchi, K. Tanaka, T. Kajiyama, A. Takahara, and S. Tasaki, *Macromolecules,* **36**, 1235 (2003).
[7] M. Doi and S. F. Edwards, *The Theory of Polymer Dynamics*, Oxford University Press, pp. 213 (1986).
[8] R. Schnell and M. Stamm, *Macromolecules*, **32**, 3420 (1999).
[9] J. M. Milhaupt, T. P. Lodge, S. D. Smith, and M. W. Hamersky, *Macromolecules*, **34**, 5561 (2001).

Chain Dimension of Polystyrene-like Hyperbranched Polymers with Various Chain End Groups in THF

Moriya Kikuchi,[1] Motoyasu Kobayashi,[1] Akihiro Tanaka,[2] and Atsushi Takahara[1]

[1]Institute for Materials Chemistry and Engineering, Kyushu University, 744 Motooka, Nishi-ku, Fukuoka, 819-0385, Japan
[2]Synthesis Research Department, Chemical Reserch Laboratories, Nissan Chemical Industries, Ltd., 722-1 tsuboi, Funabashi 274-8507, Japan

E-mail: takahara@cstf.kyushu-u.ac.jp

Abstract. A hyperbranched polymer (HPS-DC) synthesized using *N,N*-diethylamino -dithiocarbamoylstyrene as an inimer and a hyperbranched poly(4-methylstyrene) (HPS-H) prepared by hydrogenation of the HPS-DC are characterized in tetrahydrofuran (THF) solutions by static and dynamic light scattering measurements. The experimental values of z-averaged mean-square radius of gyration ($<S^2>_z^{1/2}$) and hydrodynamic radius (R_H) of the HPS-H have smaller dimensions than those of the linear polystyrene (PS) at the same degree of polymerization (N_w), whereas the R_H of the HPS-DC in the region of $N_w < 700$ is larger than that of the linear PS at the same N_w. The N_w-dependence of $<S^2>_z^{1/2}$ and R_H for the HPS-H are described by a randomly trifunctional branched polymer model with the molecular weight per a branch point (4.55×10^3 g/mol) and the chain dimensions of the linear PS in THF at 25 °C.

1. Introduction

Hyperbranched polymers are usually prepared by condensation reactions between the A and B groups using AB_f monomers ($f \geq 2$). The resulting product is a highly polydisperse mixture of randomly branched molecules. The hyperbranched polymers generally have smaller dimensions and higher density of terminal functional groups on their surfaces than those of corresponding linear polymers with the same degree of polymerization. Moreover, the hyperbranched polymers have unique physical and chemical properties such as low viscosity both in the solution and in the melt state in comparison with their linear analogues [1-3], and thus potential in variety of commercial applications.

Ishizu et al. reported synthesis of polystyrene-like hyperbranched polymer (HPS-DC) by one-pot photopolymerization of *N,N*-diethylaminodithiocarbamoylmethylstyrene, which can also work as an initiator for polymerization to afford branching point, so-called "inimer" [4]. They estimated the degree of branching (*DB*) by kinetic approach [5] and NMR [6] to be as high as *ca.* 0.3 ~ 0.4.

We have recently reported the chain dimension of the same HPS-DC in tetrahydrofuran (THF) and methyl ethyl ketone by dynamic light scattering measurement, and obtained the hydrodynamic radius (R_H) being larger than that of the linear PS with the same degree of polymerization because of the different local structure of polymer chain [7].

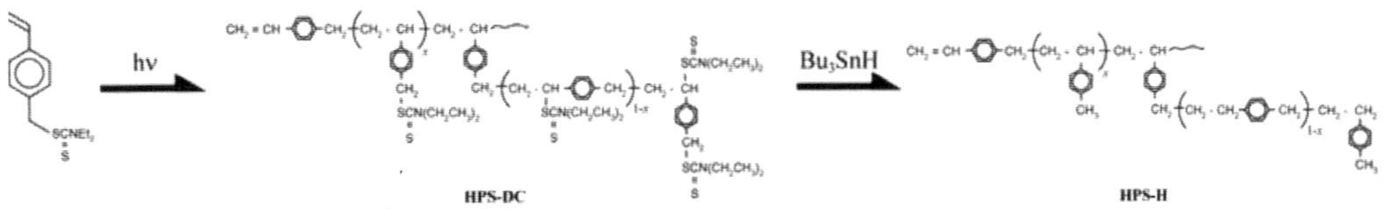

Scheme. Preparation of polystyrene-like hyperbranched polymers (HPS-DC and HPS-H).

In this study, the chain dimension of polystyrene-like hyperbranched polymers (HPS-DC and hydrogenated HPS-H) in THF were measured by static and dynamic light scattering measurements. Especially, relationship between a chain dimension and the molecular size of chain end groups were discussed.

2. Experimental

2.1. Polystyrene-like Hyperbrached Polymers

The polystyrene-like hyperbranched polystyrene (HPS-DC) was synthesized by photopolymerization of *N,N*-diethylaminodithiocarbamoylmethylstyrene as an inimer according to a previous procedure reported by Ishizu and Mori [1]. The hyperbranched poly(4-methylstyrene) (HPS-H) was prepared by hydrogenation of HPS-DC in the presence of tri-*n*-butyltin hydride. The obtained HPS-DC and HPS-H were fractionated to 11 parts with different size by a recycling preparative HPLC system (Japan Analytical Industry Co. Ltd. LC-9104) equipped with a RI detector (JAI RI-7s) and a UV-vis detector (JAI 310) using two columns (JAIGEL 3H-40 and 4H-40) in series with chloroform as the eluent at a flow rate of 14 mL/min at room temperature. The composition of styrene unit (x) of the each HPS-DC fraction was determined by ^{1}H NMR to be 0.77 $\pm$ 0.02, which was independent on the weight-averaged degree of polymerization (N_w). The degree of branching (DB) of the HPS-DC in the region of $38 < N_w < 720$ evaluated form value of x was 0.48 $\pm$ 0.02, and larger than that of reported by Ishizu et al. [5, 6].

Size exclusion chromatography (SEC) was carried out using a RI detector (SHIMADZU RID-10A) and a UV-vis detector (SHIMADZU SPD-10AV vp) equipped with three columns (TOSOH TSKgel G4000H$_{XL}$, G5000H$_{XL}$, and G6000H$_{XL}$) in THF as the eluent at a flow rate of 0.5 mL/min at 40 °C. Weight-averaged molecular weight (M_w), z-averaged mean-square radius of gyration ($\langle S^2\rangle_z$), and R_H of the fractionated HPS-DC and HPS-H were determined by the SEC equipped with a multiangle light scattering detector (MALS; Wyatt Technology DAWN-EOS, wavelength: λ = 690 nm) and a quasi-elastic light scattering detector (QELS; Wyatt Technology WYATT-QELS, scattered angle: 99.8 °) (SEC-MALS-QELS) at room temperature. The Rayleigh ratio $R(90)$ at a scattered angle of 90 ° was based on that of a pure toluene. The corrections for sensitivity of 16 detectors at angles of other than 90 ° and dead volume for each detector were performed using the scattering intensities of 0.30 wt% THF solution of a PS standard with $M_w = 2.23 \times 10^4$ g/mol and $M_w/M_n = 1.0_4$. Polymer sample solutions with mass concentration (C_p) of about 1×10^{-3} g/mL were injected using a sample loop of 100 μL to SEC columns.

The specific refractive index increment (dn/dc) of the HPS-DC and HPS-H in THF at 25 °C was measured using a differential refractometer (Otsuka Electrics DRM-1020, wavelength: λ = 633 nm). The measured dn/dc of the HPS-DC and HPS-H in THF were 0.225_9 mL/g and 0.178_8 mL/g, respectively. Cauchy dispersion formula shows that there is linear relationship between dn/dc and λ^{-2} to reveal independence of wavelength at longer region. Thus, dn/dc at λ = 633 nm was also used for the analysis of SEC-MALS-QELS.

3. Results and Discussion

Figure 1 shows a typical SEC curve of a fractionated HPS-DC ($M_w = 3.91 \times 10^4$ g/mol, $M_w/M_n = 1.1_4$), together with the absolute calibration curves of the HPS-DC (circles), HPS-H (squares), and PS (triangles) determined by SEC-MALS. It should be noted in this figure that the M_w of HPS-DC and

HPS-H in the region of all retention volume (*RV*) were larger than that of the linear PS at the same *RV*. The M_w of HPS-DC was smaller than those of the corresponding the HPS-H in the smaller *RV* region than 23 mL, whereas the M_w of HPS-DC was larger than those of the corresponding the HPS-H in the larger *RV* region than 23 mL. These results clearly demonstrate that the M_w and polydispersity index of the HPS-DC and HPS-H chains determined with a conventional PS calibration curve were significantly underestimated than those of actual HPS calculated from the corresponding absolute calibration curve determined by SEC-MALS. We supposed that the effect of branching structure of polymer chain and molecular size of chain end group appeared in the calibration curves.

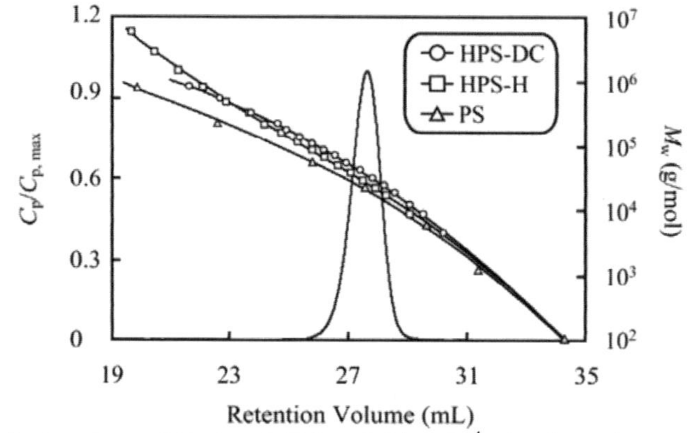

Figure 1. SEC curve of a fractionated HPS-DC ($M_w = 3.91 \times 10^4$ g/mol, $M_w/M_n = 1.1_4$) in THF, together with the absolute calibration curves of the HPS-DC (circles), HPS-H (squares), and PS (triangles) determined by SEC-MALS.

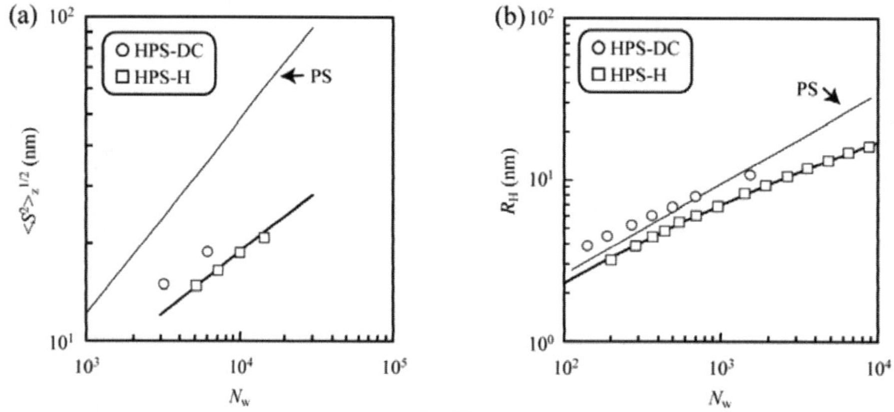

Figure 2. (a) N_w-dependence of the measured $<S^2>_z^{1/2}$ for the HPS-DC (circles) and HPS-H (squares) in THF, together with that of the linear PS in THF at 25 °C [8]. (b) N_w-dependence of the measured R_H for the HPS-DC (circles) and HPS-H (square) in THF, together with that of the linear PS in THF at 25 °C [9]. The bold lines show the calculated values of a randomly trifunctional branched polymer model calculated form equations 1 and 2 with $\gamma_M^{-1} = 4.55 \times 10^3$ g/mol and the chain dimensions of linear PS in THF at 25 °C.

Figure 2 (a) and (b) show the double-logarithmic plots of the $<S^2>_z^{1/2}$ and R_H versus N_w for the HPS-DC (circles) and HPS-H (square) in THF, together with that of the linear PS in THF at 25 °C [8, 9], respectively. It should be noted in their figures that both the $<S^2>_z^{1/2}$ and R_H of the HPS-H were much smaller than those of the linear PS and HPS-DC with the corresponding N_w. Another interesting point in Figure 2 (b) is that the R_H of the HPS-DC in the region of $N_w < 700$ was larger than that of the linear PS with the corresponding N_w, consistent with previous report [7]. Whereas, both the $<S^2>_z^{1/2}$ and R_H of the HPS-DC in the region of $N_w > 700$ have smaller dimension than those of the linear PS at the same N_w. Although the both the $<S^2>_z^{1/2}$ and R_H of the HPS-DC were larger than those of the HPS-H at the same N_w range, the power law exponent (*ca.* 0.47) of N_w-dependence of R_H of the HPS-DC was close to that (ca. 0.46) of the HPS-H. This was simply caused by difference in the size of functional group between DC-group and H-group. In general, power law relationships of the chain dimension and molecular size is mainly depending on the degree of branching, not on chemical structure. Therefore, polymer with bulky DC-groups afforded larger $<S^2>_z^{1/2}$ and R_H compared with HPS-H, but showed the same value of the power law relationships.

The dependence of $<S^2>_z^{1/2}$ and R_H on N_w for the HPS-H are precisely analyzed by the trifunctional randomly branched polymer model. We proposed the HPS-H as a randomly branched poly(4-methyl styrene) with trifunctional branching units, from composition of styrene unit (*ca.* 0.77) evaluated by ^{1}H NMR. The structure sensitive size parameters g_S [10] and g_H [11] for the randomly branched polymer chain with trifunctional branching units are expressed by

$$g_S = \frac{\langle S^2 \rangle_b}{\langle S^2 \rangle_l} = \left[\left(1 + \frac{\gamma_M M}{7} \right)^{\frac{1}{2}} + \frac{4\gamma_M M}{9\pi} \right]^{-\frac{1}{2}} \tag{1}$$

$$g_H^{-1} = \frac{R_{H,l}}{R_{H,b}} = \frac{1}{2}\Gamma\left(\frac{7}{4}\right)\left(\gamma_M M + \frac{1}{2}\right)^{\frac{1}{4}} + \frac{3}{4}\Gamma\left(\frac{5}{4}\right)\left(\gamma_M M + \frac{1}{2}\right)^{-\frac{1}{4}} \tag{2}$$

where the subscripts *b* and *l* indicate branched and linear molecules of the same degree of polymerization, γ_M^{-1}, *M*, and $\Gamma()$ are the molecular weight per a branch point, the molecular weight, and a gamma function, respectively. Figure 2 (a) and (b) show the comparison of the experimental data of the HPS-H with calculated values (bold lines) form equations 1 and 2 with the fitting parameter $\gamma_M^{-1} = 4.55 \times 10^3$ g/mol and the chain dimensions of linear PS in THF at 25 °C. Their theoretical curves in their figures were close to the experimental points. Thus, the HPS-H can be categorized as a randomly trifunctional branched poly(4-methyl styrene).

4. Conclusion

The M_w and polydispersity index of the HPSs determined with a conventional PS calibration curve are significantly underestimated than those of actual HPS calculated from the corresponding absolute calibration curve determined by SEC-MALS. The chain dimensions of the HPS-H have smaller dimensions than those of the linear PS and HPS-DC at the same N_w, whereas the R_H of the HPS-DC in the region of $N_w < 700$ is larger than that of the linear PS at the same N_w. We supposed that bulky chain end group has a strong effect on chain dimension of hyperbranched polymer with decreasing N_w. The experimental N_w-dependence of $<S^2>_z^{1/2}$ and R_H for the HPS-H in THF are described by a randomly trifunctional branched polymer model with $\gamma_M^{-1} = 4.55 \times 10^3$ g/mol and the chain dimensions of linear PS in THF at 25 °C.

References

[1] Jikei M and Kakimoto M 2001 *Prog. Polym. Sci.* **26** 1233
[2] Gao C and Yan D Y 2004 *Prog. Polym. Sci.* **29** 183
[3] Yates C R and Hayes W 2004 *Eur. Polym. J.* **40** 1257
[4] Ishizu K and Mori A 2000 *Macromol. Rapid Commun.* **21** 665
[5] Ishizu K, Ohta Y and Kawauchi S 2005 *J. Appl. Polym. Sci.* **96** 1810
[6] Ishizu K, Ohta Y and Kawauchi S 2002 *Macromolecules* **35** 3781
[7] Matsuda Y, Kobayashi M, Takahara A, Tanaka A, Hayashi H, Annaka M and Sato N 2008 *Polym. J.* **40** 375
[8] Nakamura Y, Wan Y, Mays J W, Iatrou H and Hadjichristidis N 2000 *Macromolecules* **33** 8323
[9] Fetters L J, Hadjichristidis N, Lindner J S and Mays J W 1994 *J. Phys. Chem. Ref. Data* **23** 619
[10] Zimm B H and Stockmayer W H 1949 *J. Chem. Phys.* **17** 1301
[11] Kurata M and Fukatsu M 1964 *J. Chem. Phys.* **41** 2934

AUTHOR INDEX

AUTHOR INDEX

Institute of Physics
Dirac House, Temple Back
Bristol BS1 6BE UK